PLANTS & ANIMALS

T0222208

PLANTS & ANIMALS

THE BIOLOGY SECTIONS FROM
ELEMENTARY SCIENCE. BY H. WEBB AND
M. A. GRIGG

by

M. A. GRIGG, B.Sc.

Edited by

PROFESSOR H. MUNRO FOX, F.R.S.

CAMBRIDGE
AT THE UNIVERSITY PRESS
1937

CAMBRIDGE
UNIVERSITY PRESS

University Printing House, Cambridge CB2 8BS, United Kingdom

Cambridge University Press is part of the University of Cambridge.

It furthers the University's mission by disseminating knowledge in the pursuit of
education, learning and research at the highest international levels of excellence.

www.cambridge.org
Information on this title: www.cambridge.org/9781107642799

© Cambridge University Press 1937

First published 1937
First paperback edition 2014

A catalogue record for this publication is available from the British Library

ISBN 978-1-107-64279-9 Paperback

CONTENTS

PREFACE

Many requests have been made to re-issue the Biology chapters from *Elementary Science* by H. Webb and M. A. Grigg. This has been done without any change.

This book is intended for children from 11 to 14 or 15 years old. It is written for the child and not for the teacher. Consequently the scientific terms have been omitted, so that the child can read and understand the subject.

Such a wide field has been covered that some subjects have been dealt with very briefly. The teacher should supplement these subjects, and for this reason the names of a number of reference books are given at the end of the book.

The division of work for each year is left to the teacher. When making the syllabus, however, it is necessary to remember that many specimens are only obtainable at certain times of the year. These lessons must be taken during the appropriate season.

The specimens should be provided by the children, and not by the teacher. Exceptions of course must be made in town schools. The Biology room must be a room of "Living Things".

M. A. GRIGG

KINGSTON-UPON-THAMES
1 *June* 1937

Chapter 1

PLANTS AND ANIMALS

When we speak about non-living things, we mean things that have never lived. We speak of a dead cat as being "dead", not "non-living", because it was once alive. In this Book we shall learn something about living things. The study of living things is called *biology*.

If you are asked for the names of living things, you will probably say, cows, horses, sheep, birds, fish, butterflies, etc. You may perhaps think that only those things which move from place to place are alive. This is not true; plants such as the green scum on ponds, sea weeds, moss, ferns, flowers and trees, even when they appear dead in the winter, are alive. All living things are either plants or animals.

Differences between living and non-living things

Living things have a number of characteristics, or chief features, by which we can tell them from non-living things.

1. Living things *feed*. The food taken in is changed into different substances in the animal or plant, before it actually becomes part of the body. You all know that if you roll a snow-ball down a snow-covered hillside, it increases in size by adding more snow to its surface. A child does not grow by adding pieces of bread, meat or potato on to its body, but by eating food. This food is changed in the body into muscle, nerve, bone, blood, etc.

2. All living things *breathe*, although the breathing cannot always be seen, as, for instance, in worms and plants (see Chapters 4, 7 and 10).

3. Living things can *reproduce*, that is, they produce others like themselves. A cat has kittens, and plants have seeds which

grow into new plants. If you put a number of mice together in a cage and the same number of stones and count them at the end of twelve months, you will find that you have more mice than you started with, but you have the same number of stones. This shows that living things reproduce, but non-living things cannot. All living things are produced by living things. People used to think that food went mouldy just because it was damp, and that maggots were formed by meat that was not good. Now we know that things will not go mouldy if kept airtight, and maggots are not found on meat unless we let flies lay eggs on it.

4. Living things are affected by what there is around them. All of us can see, hear, smell, taste and feel. These are called our

Fig. 1. Leaves of Clover in the day and night positions.

senses. All animals have some or all of these senses. If a mouse sees or hears a cat coming towards it, the mouse darts down its hole. If you smell some tasty food, extra juice (called saliva) comes at once into your mouth. You shudder when you drink nasty-tasting medicine, and if someone sticks a pin into your arm you immediately jump. Whatever causes any such change in animals is called a *stimulus*. Plants also change when stimulated. When a runner bean touches a stick, it twines round it. Shoots always grow towards the light, and some plants open only in the light, for example, clover leaves (Fig. 1), daisy flowers. Roots grow downwards as a result of gravity. If your mother buys flowers, or picks them out of the garden, and puts them into a vase in a hot room, they are soon wide open. Plants, then, are affected by touch, light, gravity and heat.

Differences between plants and animals

Most children think that all animals can move from place to place, while all plants are fixed in the soil. This is almost true, but there are a few tiny plants in water which move about, and there are some animals, such as the Coral (Fig. 2) and the Oyster, which do not.

There are three chief differences between animals and plants.

1. They require different kinds of *food*. Plants feed on the carbon dioxide gas in the air, with the help of the green substance in their leaves (see Chapter 4). Their roots take in water and salts, such as nitrates, sulphates, phosphates of potassium, calcium, magnesium and iron. The farmer manures his ground to replace these salts, which have been used up by the plants. These substances are changed in the plants into more complicated compounds.

Fig. 2. A piece of Red Coral showing animals partly embedded in the calcium carbonate shell.

The food of animals, however, is partly made up of very complicated compounds which they obtain from plants, either by eating plants, or by eating other animals which feed on plants. For instance, we may eat vegetables, fruits, etc., or meat, such as beef, that we get from cattle, which in turn feed on grass. If you think of all the things that you eat, you will find that we depend entirely on plants for our food. "All flesh is grass."

Since animals have to seek their food, they must move, and so they have muscles, nerves, and a brain. Since their food consists of complicated compounds they have a stomach to digest it.

2. The living parts of all animals and plants are made up of a jelly-like substance called **protoplasm**. The living substance is sometimes hardened by minerals, as, for example, in bone. A baby's bones are very soft and gristle-like, and gradually they become hardened by minerals.

In all plants and animals that we shall study (except the Amoeba) the protoplasm is divided up into very small compartments called **cells**. These can only be seen with a microscope. Plant cells differ from those of animals, as the former are surrounded by a wall of dead substance, called the cell wall. Plant cells have rigid walls, whereas animal cells do not.

3. The body of an animal usually has a **definite size and shape**, whereas a plant branches in all directions, so that its shape is constantly changing. Compare a horse with a tree.

Chapter 2

FLOWERING PLANTS

You will probably be surprised to hear that all plants do not have flowers, but if you think of a Fern, for instance, you will remember that you have never seen it flower. Later on you will learn something about the plants that have no flowers.

Most flowering plants have similar parts, **roots, stems, leaves** and **flowers**, each part having different uses for the plant.

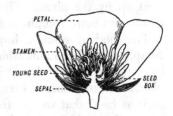

Fig. 3. Buttercup flower cut down the middle.

The flower

In the flower, seeds are produced which will later on grow into new plants. Almost all flowers have the same parts, though they may differ in colour, number and shape. If you get to know

the parts of a Buttercup you can then compare any other flower
with it.

The Buttercup

On the outside of the flower (Fig. 3) there are five green leaf-
like parts called **sepals**, which protect the flower when it is a
bud.

Inside these are five yellow leaves called **petals**, which in
some flowers are joined together to form a tube, as in the Primrose (Fig. 9). At the base of each petal of the Buttercup is

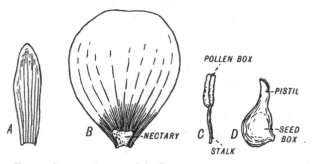

Fig. 4. Separated parts of the Buttercup flower. *A*, Sepal.
B, Petal. *C*, Stamen. *D*, Seed box.

a swelling, where a juice called **nectar** is made, which bees
turn into honey. Petals are usually brightly coloured to attract
such insects.

The **stamens** are inside the petals. Each has a stalk and a
swollen tip (the **pollen box**) where **pollen** is made (Fig. 4).
Pollen consists of fine grains like dust.

In the middle of each flower are a number of little green things
called **seed boxes**. Each contains one young seed. Some flowers,
like the Primrose (Fig. 9), have only one seed box containing
many young seeds. On top of each seed box is a **pistil**, the tip
of which is sticky when ripe. Many flowers, such as the Primrose, have long pistils.

Pollination and fertilisation

Young seeds cannot grow into new plants unless some pollen reaches them. First the pollen must reach the pistil, that is, the flower must be **pollinated**. From each grain of pollen that reaches the sticky end of the pistil, a tube grows down into the seed box. One of these tubes reaches the young seed (Fig. 5). The tube grows right into the young seed, where part of the pollen grain joins with part of the seed. This is called **fertilisation**, and can be compared with fertilisation in the Hydra or the Earthworm (pp. 65 and 68).

Fig. 5. A pollen grain fertilising a young seed.

Usually a flower is pollinated with the pollen from another flower, carried there by insects or by the wind. This is called **cross-pollination**. If this does not happen, many flowers are **self-pollinated**, that is, they pollinate themselves. Cross-pollination usually results in better seeds than self-pollination, and it is therefore to the advantage of a plant to be cross-pollinated.

Pollination by insects

Flowers pollinated by insects usually have bright colours and scent to attract these animals. The insects really come for the nectar, which is their food, and which they suck up with their "tongues". While they are collecting the nectar, pollen sticks to their hairy bodies. Later on, this comes off on the pistil tips of other flowers which they visit. Besides nectar, bees also gather pollen for their own use and carry it away as a lump stuck to each hind leg.

Many flowers provide a landing stage for the insects which visit them, for example the Dead-nettle, Snapdragon, Pea, Broom and Iris.

The Dead-nettle

This flower has five sepals, five petals joined together, one forming the landing stage and two others the "hood" (Fig. 6). Underneath the hood are four stamens, and a two-lobed pistil which sticks out below them.

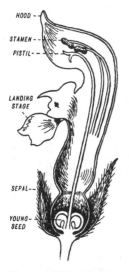

HOOD --

STAMEN --

PISTIL --

LANDING STAGE

SEPAL --

YOUNG- SEED

Fig. 6.   White Dead-nettle cut vertically.

Fig. 7. Flower head of a Daisy cut in two. The central flowers have short petals, those at the edge have three united petals drawn out into tongues.

A bee alights on the landing stage and pokes its head into the flower to get the nectar which is at the bottom of the tube of petals. The result is that pollen is first rubbed off the bee's back on to the sticky tip of the pistil. Next the stamens deposit their pollen on to the bee's back. This pollen is then taken by the bee to another flower. This is how the Dead-nettle is cross-pollinated.

The Dandelion

The Dandelion is not one flower at all but hundreds of tiny flowers growing together and forming the "head". This is also true of the Daisy (Fig. 7), Sunflower and Marigold.

There are two kinds of flowers in the "head". The outside flowers have five petals joined together to look like one petal. These flowers have no stamens.

The inner flowers have sepals which are like hairs. They have five petals joined together, five stamens joined together by their pollen boxes, and one pistil whose tip is split in two when ripe.

The stamens ripen before the pistil, and shed their pollen inwards. This pollen is pushed out by the pistil as the latter grows. When the pistil is ripe, the two halves of its tip separate and become sticky. They are then ready to receive pollen brought from another flower by insects which crawl over the "head". If the flowers are not cross-pollinated by such insect visits, the halves of each pistil tip curve backwards and pick up pollen from their own styles. [The style joins the tip of the pistil to the seed box (Fig. 8).]

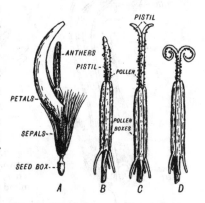

Fig. 8. Dandelion. *A*, Young flower. *B–D*, Stages in pollination.

Thus if the Dandelion cannot secure the advantage of cross-pollination with the help of insects, it falls back on self-pollination.

The Primrose

The primrose has a wonderful arrangement for cross-pollination. There are two different kinds of flowers, **pin-eyed** and **thrum-eyed** (Fig. 9), each growing on different plants. In the pin-eyed flowers the tip of the pistil is at the top of the tube made by the petals, with the pollen boxes half-way down. In the thrum-eyed flower the opposite is true.

If an insect visits a pin-eyed flower to get the nectar from the bottom of the tube, pollen from the pollen boxes half-way down the tube gets stuck to the insect's head. When next the insect goes to a thrum-eyed flower, the pollen on its head is rubbed off on to the pistil of this flower. At the same time pollen from the

Fig. 9. Primrose. Pin-eyed (*A*) and thrum-eyed (*B*) flowers cut down the middle.

pollen boxes of the thrum-eyed flower, which are at the top of the tube, gets on to the insect's body. This pollen is afterwards brushed on to the pistil of another pin-eyed flower.

The Willow

Willow flowers are also pollinated by insects. They have no brightly coloured petals, but they produce nectar to attract the insects. The pistils and stamens are found in different flowers growing on separate trees, which may be far apart. On each tree the flowers grow together in **catkins**, the "Golden Willow" having flowers with stamens, the "Pussy" or "Silver Willow" having flowers with pistils only. Look at single flowers of each kind, and compare them with Fig. 10.

Wind pollination

Flowers pollinated by wind do not need bright colours or scent, so that petals are often absent.

The stamens usually have long stalks, so that they readily blow about in the wind and scatter their pollen, which is very light.

Pistils also are long and hang out of the flowers to catch the pollen.

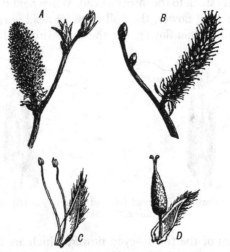

Fig. 10. Willow. *A*, Catkin of flowers bearing stamens. *B*, Catkin of flowers bearing seed boxes. *C*, Flower with stamens. *D*, Flower with seed box. Each flower has a scale-like leaf to protect it.

Fig. 11. Grass flowers with hanging pollen boxes and feathery pistils. Pollen is being blown from the right-hand flower on to the pistils of the left-hand flower.

Grass

Grass flowers are pollinated by wind. They are very small, so you must look at them through a lens. They have three hanging pollen boxes, and two feathery pistils at the top of the seed box, which catch the pollen as it is blown about. There are no petals or sepals, only scales which look like tiny leaves (Fig. 11).

Hazel

In the Hazel the pistils and stamens are found in different flowers on the same tree (Fig. 12), not on different trees as in the Willow. The flowers containing the stamens grow in catkins. Most people are familiar with the Hazel catkins, which when shaken give off a cloud of pollen.

On the same twigs you will see some green buds with red threads (really pistils) coming out of their tips. These buds contain the flowers with the pistils. Pull apart a single flower of each and compare them with Fig. 12.

When the wind blows, the catkins are shaken and pollen is blown about. Some of it will fall on these long red pistils.

Seeds and fruits

After the young seeds have been fertilised, the sepals, petals and stamens of the flower die. The seed boxes grow larger and larger and become the *fruits*, inside which are the **seeds**. Most people think a fruit is something juicy that we can eat, such as Plums, Apricots, Currants, Oranges, Melons. (A Marrow is also a juicy fruit although it is eaten as a vegetable.) The word "fruit", however, means something more than this. Many flowers form hard fruits, like the nuts of the Hazel (Fig. 12) and acorns. Pods, like those of Peas, Beans and Vetches (Fig. 15), are also fruits, but they are dry, not juicy. Inside these dry fruits are seeds.

Seed dispersal

When the seeds are ripe they must be scattered away from the parent plant. If not they will just fall on the ground beneath. In this case hundreds of young plants, called **seedlings**, would grow so closely together, that they would not be able to feed, breathe or grow, and so would die.

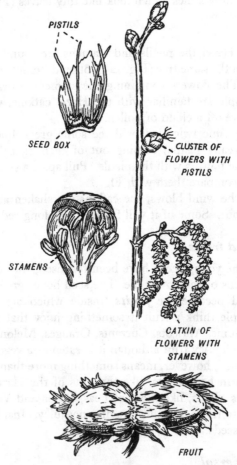

Fig. 12. Hazel.

So seeds must be scattered. *Wind*, *animals* and *water* help to carry fruits and seeds away from the parent plant. In the case of some dry fruits, the seeds are shot out when the seed box bursts; these are called ***explosive fruits***.

Wind

Fruits scattered by wind are dry and very light. They either have "wings" (Sycamore, Lime, Fig. 13) or hairs (Dandelion, Fig. 13) attached to them, so that they easily float about in the wind. You have all blown at a Dandelion "head" to tell the time, so you know how well these fruits can float.

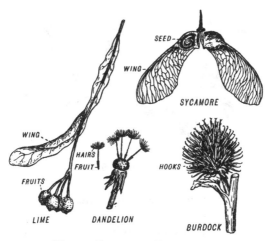

Fig. 13. Dry, non-splitting fruits.

Animals

Animals help in many ways to scatter seeds.

On the dry fruits of the Burdock (Fig. 13) are hooks, which cling to the coat of any passing animal, for instance a sheep, that touches them.

Squirrels whilst gathering and storing nuts for the winter may scatter them.

Some birds eat juicy berries like those on Hawthorn and Holly trees (Fig. 14). They eat the juicy part and drop the seed.

It is interesting to know how birds scatter Mistletoe seeds. When a bird pecks one of the white berries, which are sticky, the berry sticks to the bird's beak. The bird cannot get rid of the seeds, so rubs its beak on the branch of a tree, thus rubbing off the seeds, which then begin growing on the branch. The young Mistletoe plant has no roots, so it sends a sucker into the branch to get some of its food. Since it lives thus on another living plant the Mistletoe is a *parasite*. Later you will read of an animal parasite, the Threadworm (p. 69).

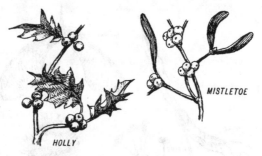

Fig. 14. Juicy fruits.

Mud containing small seeds often sticks to the feet of birds, and is carried long distances.

Seeds of corn and edible fruit trees are, of course, often taken long distances by man and planted in other parts of the earth.

Water

Plants living in or near moving water have fruits which float. These may drop or be blown into the water, and may then be carried to other banks or shores. Coconut Palms are found near the sea shore in hot countries. (The coconuts bought in shops are not complete, as the outer part which makes them float has been cut away.)

Explosive fruits

When the fruit is a dry one, its walls may split and shoot out the seeds, while the fruit is still joined to the plant. In the Gorse, Broom and Vetch (Fig. 15), for example, the pods dry and split in two. Each half curls up, forcing out the seeds some feet away. Sometimes the popping of the Gorse can be heard as the pods burst violently.

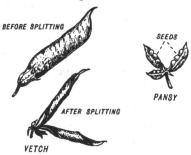

The seed box of the Pansy is divided into three boat-shaped parts, which separate when the fruit is ripe (Fig. 15). The sides of each part in drying press on the seeds and shoot them out.

Fig. 15. Dry, splitting fruits.

Structure of the Bean seed

Before looking at a Bean seed, soak it for about a day in water; then it will be larger and softer.

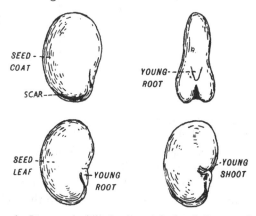

Fig. 16. Bean seeds. The bottom right-hand diagram shows the inside of the seed with one seed leaf removed.

A broad *scar* is seen at the thick end of the seed, where the seed was joined by a stalk to the pod (Fig. 16). Covering the seed is a thin *seed coat*, which can be removed.

Inside the seed coat are two large fleshy *seed leaves*, which can easily be separated. Where they are joined together there is a small pointed *young root*, which can be seen through the seed coat, before it is removed. Between the seed leaves is a small bud, the *young shoot*, consisting of a very small stem and folded leaves.

Germination

Usually a period of rest is necessary after seeds are formed before they will sprout or germinate. Most seeds wait through

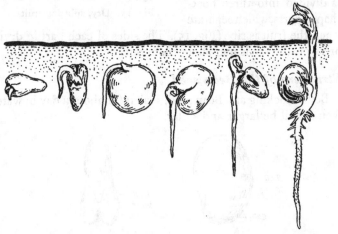

Fig. 17. Bean seed germinating. The young root grows downwards, and the young shoot upwards whatever position the seed may be in.

the winter and germinate in the spring. Many seeds are able to wait years before they sprout. Wheat and Barley will wait for ten or more years. Others, like the Poplar, only wait for a few weeks.

To find out how seeds grow we must watch them closely day by day. The easiest way to do this is to take a glass jar and line it with blotting paper. Moisten the blotting paper and place the seeds half-way down the jar, between the glass and the blotting paper. Seeds often fall to the bottom of the jar if the blotting paper is not moistened before the seeds are put in the jar. Wet the blotting paper from time to time.

Germination of the Bean seed

When the Bean germinates, the seed coat splits and the young root grows downwards, no matter in which direction the seed may be lying (Fig. 17). Next the young shoot grows out, and then grows upwards. It is bent double as it pushes its way through the soil, so that the young leaves are not injured. When the young stem reaches the top of the soil, it straightens and its leaves turn green. Meanwhile the root has become longer and side roots have grown from it. All this time the young plant has been using the food in the seed leaves, which slowly shrink in size.

Germination of the Marrow seed

When the Marrow seed germinates, the seed coat splits along its edges (Fig. 18) and the young root grows downwards. The root gradually grows longer and side roots grow from it. Meanwhile the young shoot grows upwards, bent as in the Bean,

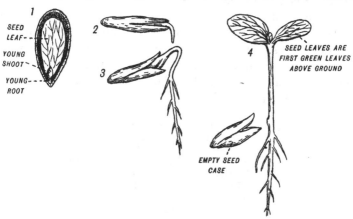

Fig. 18. Germination of Marrow seed.

drawing the seed leaves with it. The seed leaves turn into the first green leaves above the ground. The empty seed case is all that remains of the seed in the soil.

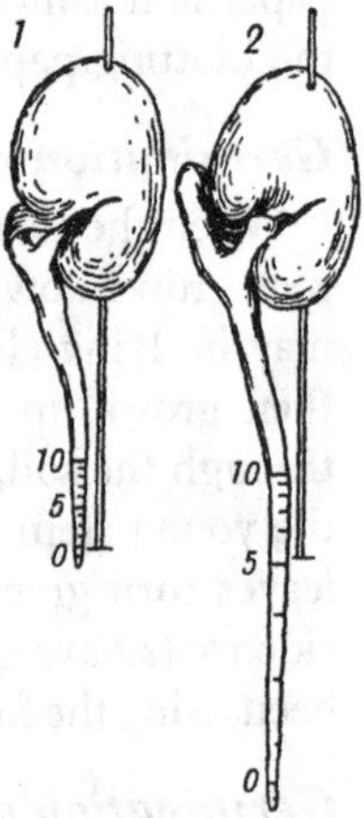

Conditions necessary for germination

Under certain conditions seeds will not germinate very well, and may not even germinate at all (see Expt. 12, Chapter 2).

During the winter, seeds will not grow because it is too cold. They begin to sprout when the warmer days of spring arrive. Seeds will not grow at a very high temperature either, but in this country such a temperature is seldom reached under natural conditions.

Water and oxygen are also necessary for germination, but most seeds germinate equally well in the light or in the dark (Expt. 12, Chapter 2).

Fig. 19. Germinating Bean. 1. Root tip marked with ink lines 1 millimetre apart; 2. Same root 24 hours later.

Region of growth of root and stem

As the root and shoot grow longer, the part behind the tip grows more quickly than the older part. This can be shown by marking lines, with cotton dipped in Indian ink, 1 millimetre apart, from the tip along the root of a seedling. After several days the marks immediately behind the tip are much farther apart than the others (Fig. 19).

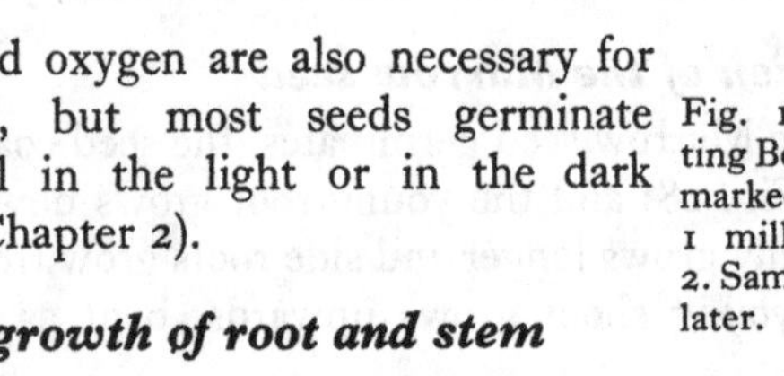

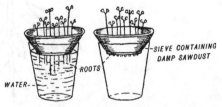

Fig. 20. Experiment to show that roots grow towards water.

Conditions affecting growth

TEMPERATURE has the same effect on growth as on germination.

GRAVITY. The fact that roots grow downwards is due to the force of gravity.

LIGHT. Seeds grown in the dark at first grow as well as those in the light (see Experiments). After a time, however, you will see that the stems of plants growing in the dark are very long and weak. The leaves are yellow, as no green colour can be formed without light. Shoots always grow towards the light (see Experiments and Fig. 21).

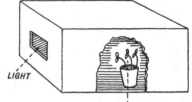

SEEDLINGS GROWING TOWARDS THE LIGHT

Fig. 21. Effect of light on direction of growth in plants.

WATER has a stronger effect on roots than gravity. Roots in the earth will turn from a dry part of the soil to one that is wet, even if they have to grow sideways (see Experiments and Fig. 20).

Chapter 3

FLOWERING PLANTS [*contd.*]

The structure of roots

After having learned many things about flowers, fruits and seeds, we must now study the roots of plants. The root is the first part of the plant to grow out of any seed. If we watch a Bean seed sprouting, we notice that the young root grows downwards for about one and a half inches. From this root, which is called the **primary root**, side or **lateral roots** then begin to grow (Fig. 22).

We have already found that the growing region of a root is

immediately behind the tip. This part is very delicate and must be protected from harm while it is pushing its way through the soil, so the growing region is covered with a **root cap**. Just

behind the tip of all roots there are things looking like hairs which are single living cells. These cells, which are called **root hairs**, are more numerous on roots exposed to moisture than they are on roots in dry soil. The work of the root hairs is to take in water and food. This food consists of salts in solution. The root hairs cling to the particles of soil and can thus make use of the thin film of moisture around them (Fig. 23).

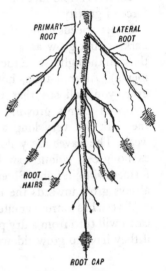

Fig. 22. The roots of a plant.

Kinds of roots

Any primary root which grows much larger than its branches is called a **tap root**. In some plants the tap root is used as a storehouse for food, as it is in the Carrot and Turnip (Fig. 24). When you watched the Maize grain sprouting, you noticed that other roots, as well as the primary root, grew out of the grain, and these soon equalled the primary root in length (Fig. 25). A plant such as a Grass (Fig. 24 *C*) which has no main root is said to have a **fibrous root**. As there is no primary root to anchor the plant in the soil, the fibrous roots are spread out to fix the plant firmly in the ground.

Roots, such as those of the Maize, which do not arise from the primary root are called **adventitious roots**. These adventitious roots often grow out of stems. For example, roots often come out of the cut ends of stems (Expt. 7, Chapter 3) or out of creeping stems (p. 25). The Banyan trees of tropical forests

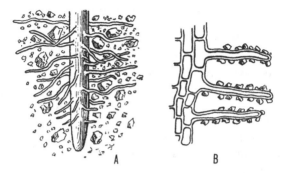

Fig. 23. Root hairs. *A* shows root hairs growing out from behind the tip of a root and penetrating the soil. *B* shows three root hairs more magnified. Each root hair is seen to be one cell. Soil particles are sticking to the root hairs.

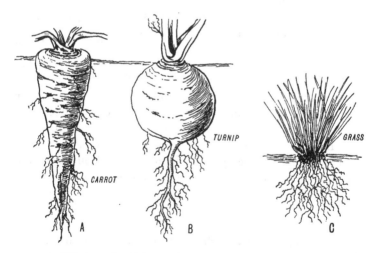

Fig. 24. *A* and *B* are tap roots. *C* is a fibrous root.

have strong woody roots growing down from their branches, and these make the undergrowth very dense.

Uses of roots

We have already learned that roots of plants have several functions.

1. They fix the plant firmly to the soil.

2. They obtain water and salts in solution for the plant.

3. Sometimes food is stored in the primary root.

Roots may have special functions. The Ivy, for example, has very short roots growing out of its stems. These roots cling to any near object, so giving support to the plant. The next time that you see Ivy growing up a tree or up the wall of a house, look for these roots, and try to pull them away from their support.

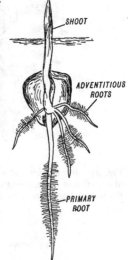

Fig. 25. Maize grain showing the primary root and adventitious roots growing out of the grain, which will soon equal the primary root in length.

The stem

After the roots have begun to grow from the seed, the young **stem** grows out. From the stem side outgrowths called **leaves** grow, which are different in appearance from the stem. A root also has side outgrowths, but these are always like the root. Because of this it is easy to distinguish roots from stems even though the stem may be underground as in the Buttercup (Fig. 27 *B*).

Annuals

The structure of stems depends upon the work they have to do. Some plants, for example the Sweet Pea, germinate from the seed and then in a single year grow into a plant producing flowers and seeds. Such plants are called **annuals**. In these plants the stems have to grow very quickly to carry the leaves and flowers to the light. The stems also remain green and can do the same work as the leaves. You will learn later what work the leaves do for the plant. It is not necessary for the stems of annuals to be thick, or well protected, as they have few flowers and leaves to bear, and will die before the winter comes.

Biennials and Perennials

All plants are not annuals. The Carrot, Parsnip, Beetroot and Radish, for example, live for two years. During the first year roots and leaves grow and food is stored in the primary root. In the second year the stored food is used as the flowers and seeds grow. Then the plant dies. These plants are called **biennials**. Plants like trees, that live for more than two years, are called **perennials**. The stems of perennials have to exist through many winters, so they are protected by a dark, tough and thick covering which is made of cork.

A plant with stems which are soft and short-lived is called a **herb**. Herbs are usually either annuals or biennials, although sometimes they may live for several years. The Chrysanthemum is an example of a perennial herb. A **tree** has a distinct main trunk which bears branches. A **shrub** is usually smaller than a tree and has several branches which arise together from one point as in the Elderberry. Trees and shrubs have woody stems and are perennials.

Growth in thickness of stems

As trees and shrubs get older, their stems have to supply the ever-increasing number of leaves with water and salts in solution. These have been taken in by the roots. Also the stem has to bear the weight of more and more branches and leaves. To do this the stems increase in width. The growing and dividing cells which add to the width form part of the greenish layer immediately under the **bark**. The bark itself is a dead covering of cork which prevents loss of water. As the trunk increases in thickness the bark becomes too small and it splits. New layers of cork are then formed beneath the old ones. In the Oak and Elm you will see very deep cracks in the bark. Other trees, such as the Plane and Birch, have smooth bark which comes off in flakes.

As the cells underneath the bark divide, those on the inside are changed into cells with hard, thick cell walls which gradually die. These are called **wood cells**. Water, with the salts in solution, only passes up through those wood cells just inside the bark which are alive. These cells form the **sap wood**. As new sap wood is formed, the older sap wood gradually dies forming the central dry

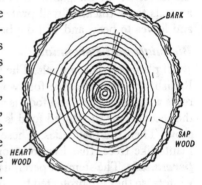

Fig. 26. Transverse section of a branch showing annual rings.

mass of dead wood cells, which is called the **heart wood**. The heart wood is the chief support or skeleton of the trunk.

The cells of the wood which are formed in the spring are larger than those formed in the autumn when growth is slow owing to the low temperature. So a clear ring of wood is added each year. These rings are called **annual rings**. They can easily be seen in a section of a tree (Fig. 26). By counting the number of rings you can tell the age of a tree. Some giant trees are 1000 years old.

Knots in wood are sections across the bases of branches which grew out in a certain year. In succeeding years these bases have become enclosed by the growth in thickness of the stem.

Stems with special functions

Normal stems grow above ground but there are some stems which live in the soil and have special functions. In Chapter 2 we learned that the normal method of reproduction in flowering plants is by seeds. Some plants, however, use stems for this purpose, and so produce new plants more quickly. There are several types of these underground stems.

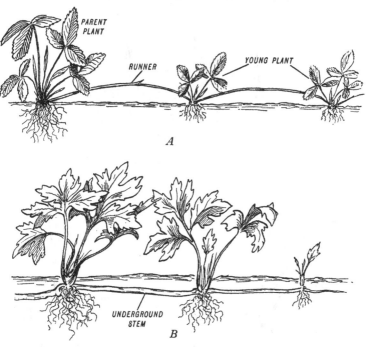

Fig. 27. Runners. *A*, Strawberry. *B*, Buttercup.

*1. **Runners***. These are really creeping stems, which grow underground in the Buttercup, but above ground in the Strawberry (Fig. 27). Each one has a bud at its tip which can produce a new plant; roots, which are adventitious, grow out from its base. The creeping stem then dies and the new plant is free.

*2. **Rhizomes***. Couch grass (Fig. 28) and Iris have long running underground stems. At intervals along these stems,

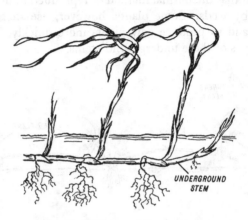

Fig. 28. Rhizome of Couch Grass.

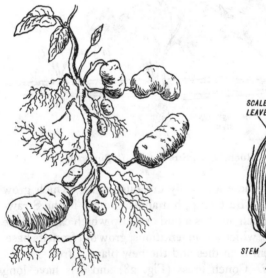

Fig. 29. Tubers of Potato plant.

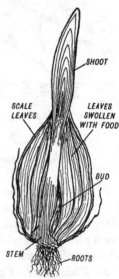

Fig. 30. Bulb cut in two.

adventitious roots and leaves arise. The stem is terminated by a bud. Couch Grass is a troublesome weed in the garden as these underground stems often stretch for several feet, and after weeding, the plant will appear again unless every trace of the stem is dug out.

3. Tubers. The Potatoes which we eat are not roots, as so many people think, but swollen stems, or tubers, where food is

Fig. 31. Climbers. *A*, Hop twining in same direction as the clock. *B*, Convolvulus twining in the opposite direction.

stored. The eyes are buds, which, when the tuber is buried in the soil, develop into long thin underground stems that bear more tubers (Fig. 29). Adventitious roots also grow from these places on the Potato, but the roots themselves never bear tubers. As the Potatoes are stems, not roots, they will turn green when growing if they are not covered with soil.

4. Bulbs. If you cut an Onion in two lengthwise you will see its structure (Fig. 30). At the base of the bulb there is a small pointed stem from the under side of which roots arise. The remainder of the bulb consists of leaves which are swollen with food. These leaves are protected on the outside by other scale-like leaves. The way in which these leaves are arranged may be compared with the arrangement of leaves in an ordinary bud. A bulb is really a large bud, whose leaves are swollen with food. A new bulb arises as a small bud on the short stem at the base of one of the fleshy leaves (Fig. 30).

Climbing stems

The stems of some plants are too weak to support the weight of the leaves and flowers, so they climb up any plant or obstacle near them. When we grow Kidney Beans or Hops we always put sticks near the young plants, up which they may climb. The stems twine themselves round these supports. The stems of the Hop and Honeysuckle revolve in a clockwise manner, whereas those of the Kidney Bean and Convolvulus go round in the opposite direction (Fig. 31).

Chapter 4

FLOWERING PLANTS [contd.]

To complete our study of Flowering Plants we have still to learn about the structure and uses of the leaves. Let us first consider the leaves when they are very small.

Buds

Look at the twigs of any tree that loses its leaves at the beginning of winter, and you will see a number of **buds** (Fig. 32). A bud is a short shoot with a number of young **leaves** on it closely packed together (Expt. 1, Chap. 4). At the end of every twig there is a bud, called the **terminal bud**. In addition, there is a bud at the base of every leaf on the plant, between the leaf stalk and the stem of the twig. These buds are called **axillary buds** because they grow in the **axils** of the leaves (Fig. 32).

The young leaves inside the buds are extremely delicate. When they are first forming they require protection from too great heat, cold, dryness, and moisture, as well as against the attacks of animals. Buds are protected in various ways. They are often protected by the stalk of the leaf until it falls off the tree. In the Plane each bud is completely covered by the hollow end of the stalk (Fig. 33).

Horsechestnut buds are large, dark-coloured and sticky (Fig. 32). This gum which makes the buds sticky is only on the outer surface of the bud and protects the young leaves from insects and other animals. Carefully pull a Horsechestnut bud to pieces. On the outside are small leaves called **scale leaves**, which are packed closely together and are waterproof. The spaces between the young leaves themselves are filled with hairs which protect the young leaves from the cold, and also prevent them from losing too much moisture (see Transpiration) and so drying up.

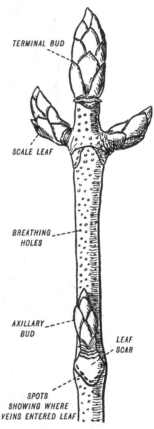

Fig. 32.
A Horsechestnut twig.

Cut a bud in two lengthwise and you will see a very short stem with many small leaves growing out of it (Fig. 34 *A*). In the spring the end of the stem, in the middle of the bud, will

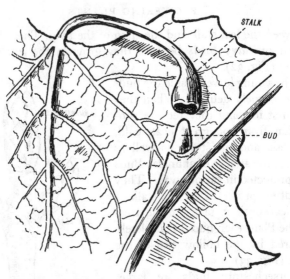

Fig. 33. Plane showing bud protected by leaf stalk.

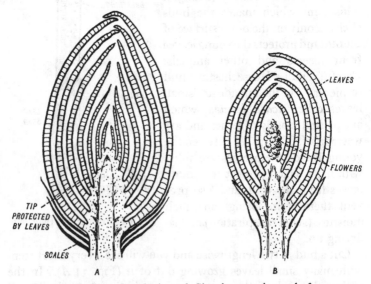

Fig. 34. Buds cut lengthwise. *A*, Showing growing end of stem. *B*, Showing young flowers.

lengthen, and the green leaves will expand and grow (Expt. 2, Chap. 4). No cap protects the growing end of the stem as it did in the root, because the stem does not have to push its way through anything hard such as the soil. The delicate growing end of the stem is well enough protected by the young leaves which are neatly folded over it (Fig. 34 *A*).

Some of the terminal buds, when cut open, show the beginnings of a flower as well as leaves (Fig. 34 *B*). The warmth of spring causes the buds to open. Since the leaves and flowers are already formed inside the bud, they soon grow out.

Twigs

Twigs are small branches of trees. Look at a Horsechestnut twig during the winter (Fig. 32). Beneath each bud there is a horse-shoe-shaped scar, which shows where a previous year's leaf was joined to the stem. These scars are called **leaf scars** (see Leaf fall). On the scar there are a number of small spots, usually seven in the Horsechestnut, which mark the places where veins from the stem entered the leaf.

If a terminal bud grows into a short stem with flowers at the end of it, it stops growing. Later on the flowers fade and the short stem which bore them drops off, leaving a scar at the end of the stem. The following year the growth of that twig is continued by the next pair of axillary buds. This makes a fork, in the middle of which is the scar of the flower stem (Fig. 35).

On the thin bark you will see many small spots. These are **breathing holes**. The bark is a dead covering, but the living cells underneath must breathe. This they do through these holes.

Keep some twigs in water in a warm room and watch the buds opening (Expt. 2, Chap. 4). You will notice that the scale leaves fall off as the bud opens, leaving a ring of scars on the stem. The ring of scars made by the scale leaves of the last year's terminal bud can be seen lower down the stem. In one year the stem has grown from this ring of scars to the terminal bud. As these rings denote a year's growth we can call them **growth rings**. By counting the number of rings the age of a twig is found. Twigs vary very much in the amount they grow in one year. Some twigs grow several inches, or even several feet, as for example Lime trees. Others may grow only a fraction of an inch in one

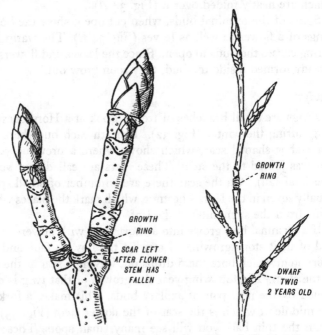

Fig. 35. Horsechestnut showing fork made by growth of axillary buds after the flower stem has fallen.

Fig. 36. Beech twig.

year. Look at the small side branches on a Beech twig (Fig. 36). Although very short they are probably several years old. These small branches are called **dwarf twigs**.

Side branches usually grow more slowly than the main branch, but if the main branch is cut, the side branches grow more quickly. If you have a hedge of Privet, or any other plant, you must repeatedly cut off the terminal branches so that the side branches will grow more quickly and produce a thicker hedge.

The recognition of common twigs

Trees can be recognised during the winter by their twigs. The bark of twigs varies in colour and the buds differ in size, shape,

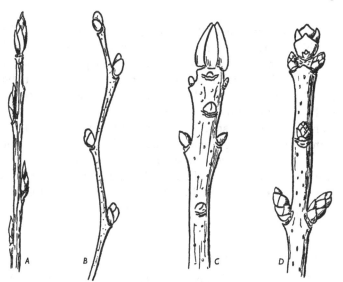

Fig. 37. Twigs. *A*, Poplar. *B*, Lime. *C*, Ash. *D*, Sycamore.
For description see below.

colour and their arrangement on the stem (Expt. 4, Chap. 4). Twigs of six of the most common trees are briefly described here (Fig. 37).

Poplar. Pointed buds, arranged alternately, and lying close to the light brown stem.

Lime. Reddish-green stems and buds. Round buds arranged alternately down "zig-zag" stem.

Sycamore. Greyish stems. Green buds in pairs down the stem. One terminal bud.

Ash. Grey stems. Black buds.

Beech. Light-coloured bark. Long, thin, pointed, brown buds arranged alternately. Dwarf twigs (Fig. 36).

Horsechestnut. Large, sticky, brown buds arranged in pairs down the stem. Horse-shoe-shaped scars (Fig. 32).

Leaves

It is necessary for all leaves to get as much air and light as possible. Leaves are arranged on the stem so that one leaf does not shade another. Sometimes they are in pairs opposite one another, and each pair is usually at right angles to the pair above, as in the Privet (Fig. 38 *A*). In most plants the leaves come off singly, as for example in the Hazel (Fig. 39). In spite of their arrangement on the stem, the leaves on the branches at the side of the tree cannot get enough light. So the leaf stalks often twist to bring the leaves into such a position that they receive the greatest amount of light (Fig. 38 *B*). The leaves of a Dandelion have no long stalks to hold them out to the light. Instead its leaves spread out to form a rosette so that each leaf gets a share of the light.

The structure of a leaf

A leaf consists of two parts: (1) The **leaf stalk**, (2) the **blade**.

The leaf stalk varies in length. Leaves growing in shady places have long stalks to hold them to the light, whereas others like the Dandelion have no stalks. The base of the leaf stalk is sometimes flattened and may even surround the stem to which it is attached.

The leaf blade varies very much in structure. A number of lines can be seen on the blade. These lines are called **veins** and are the canals along which water and food pass to the leaf, and food made in the leaf passes back into the stem. The veins also give the blade support and make it rigid.

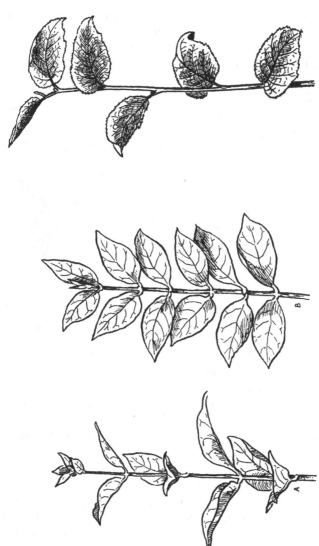

Fig. 39. Hazel branch. Leaves arranged singly along the stem.

Fig. 38. Privet branches. A, Normal arrangement of leaves in twos alternately down the stem. B, A similar twig from the side of the hedge: the leaf stalks have twisted so that the leaves can get the greatest amount of light.

Compare the leaf of a Grass or Lily-of-the-Valley (Fig. 40) with that of a Birch (Fig. 41). You will see that the veins are very differently arranged in each. In the Birch leaf there is one main vein called the *midrib* from which small veins branch forming a network over the blade. These leaves are said to be *net veined*. The veins in a Lily-of-the-Valley leaf run parallel to one another and do not form a network. This leaf is *parallel veined*.

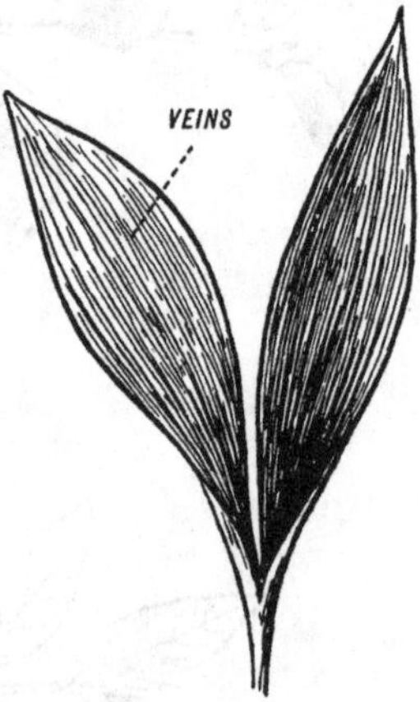

The shape of a blade varies very much (Expt. 5, Chap. 4). The edge or margin of the leaf also varies. The margin of a Privet leaf is smooth (Fig. 38), but a Birch leaf has toothed edges (Fig. 41), and an Oak leaf a wavy outline.

The blade is sometimes cut up (Fig. 42), and is not whole as it is in the Privet. The leaves of the Water Crowfoot are interesting to look at (Fig. 43). The leaves above water are

VEINS

Fig. 40. Lily-of-the-valley leaves showing parallel arrangement of veins.

very little cut up, but the leaves in the water are cut up so much that they look like bunches of green threads. This prevents them from getting damaged in the water.

As long as the divisions do not reach to the midrib, the leaf is *simple*. When the blade is divided to the midrib a number of separate *leaflets* are formed, and the leaf is then *compound* (Fig. 44).

The modification of leaves

The leaves or leaflets of some plants do not grow as ordinary leaves but form *tendrils*. These tendrils are special organs with which the plant climbs (Expt. 6, Chap. 4). We learned in Chapter 3 that the stems of some plants are too weak to support the leaves and flowers and hold the leaves out to the light.

Fig. 41. Net-veined leaf of Birch.

Fig. 42. Leaf of Blue Meadow Crane's-bill. The blade is very deeply cut up.

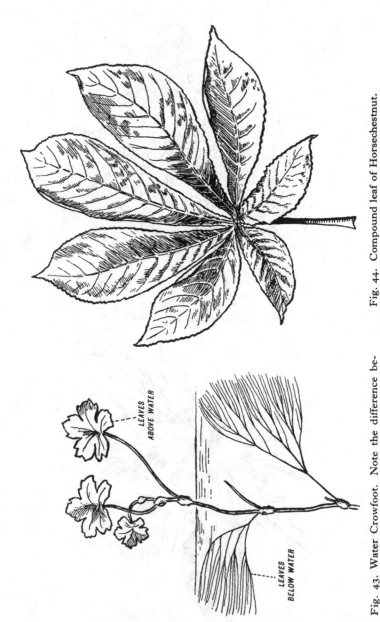

LEAVES
ABOVE WATER

LEAVES
BELOW WATER

Fig. 43. Water Crowfoot. Note the difference between the leaves above and the leaves below the water.

Fig. 44. Compound leaf of Horsechestnut.

The Hop and Convolvulus climb by twining their stems round a support. The Ivy climbs by means of the roots on its stems. The Rose climbs or scrambles with the help of its thorns.

Fig. 45. Garden Pea. Compound leaf with tendrils.

In the Pea (Fig. 45) the top leaflet of each leaf forms a tendril. The tendrils cling to any near object and so help to support the plant. In the garden we put sticks to which these tendrils can cling.

The Virginia Creeper or Wild Vine has tendrils, but these are really shoots not leaves.

The leaf blade is covered with a transparent skin which is

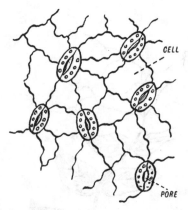

Fig. 46. The skin of a leaf seen with a microscope.

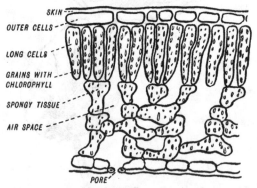

Fig. 47. A section of a leaf.

perforated, chiefly on the lower side, by many small holes or **pores** (Expt. 7, Chap. 4). These holes are very small and can only be seen with a microscope (Fig. 46). Through these holes water escapes from the leaf as water vapour, the plant breathes, and also gases which form part of the food of the plant are taken in, and unwanted gases are passed out.

Look at the diagram of a thin section of a leaf. You will see that it is made of a number of cells of different shapes (Fig. 47). The outer cells have thick walls on the outside. On the upper side of the leaf, beneath the outer cells, are long cells packed closely together. Most of the green colouring matter, called *chlorophyll*, is contained in small grains in these cells. The lower cells contain little chlorophyll, so that the lower surface of a leaf is lighter in colour than the upper surface. The cells in the lower part of the leaf form a sort of sponge with air spaces between them. These air spaces form passages between the cells of the upper surface and the pores.

Transpiration

We know that the roots of plants take in large quantities of water in which salts required by the plant are dissolved (Fig. 48. Expt. 8). What happens to all this water? Only a little is used by the plant. Most of the water goes out as water vapour through the pores of the leaves. We say that leaves *transpire*. At the end of the book a number of experiments are given to show that leaves transpire.

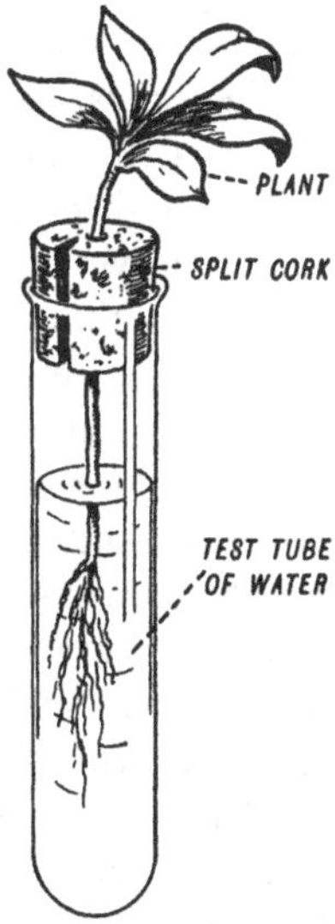

Just as we perspire when we are hot and this makes us cooler, so the evaporation of water from the leaves keeps them cool in the heat of the sun. The water lost from the leaves is replaced by more water which is drawn up through the stems from the roots. It is surprising to what heights water can be drawn upwards by the transpiration from the leaves, when you think how tall some trees are.

Fig. 48. Expt. 8. To show that roots take in water.

The amount of water transpired depends on several things.

More water is given out in the daytime, when the atmosphere is warmer, than at night.

The amount of moisture in the atmosphere also determines the amount of water vapour given out. More water vapour is given out when the air is warmer and drier. Although the tropical forests are the hottest parts of the world, the atmosphere there is very wet, so transpiration is fairly slow.

Wind also increases transpiration. The water vapour given out is quickly blown away, and the leaf is again surrounded by drier air. Plants in draughty places transpire more quickly than those in sheltered spots.

There are two ways in which plants can keep stiff. We have already learned that large plants have a skeleton of wood. Small plants have no such skeleton and are kept stiff by the *turgid* state of their cells. A cell is turgid when it is swollen out with water. When a cut plant loses water by transpiration, the cells become flabby, and the plant no longer has any support so it withers.

The leaves must not give out more water than the roots can take in, otherwise the plant will wither and may even die. Plants have several ways of preventing their leaves from losing too much water. If the air is very dry the pores close so that no water vapour can be given out. Some leaves have a very thick covering and fewer pores from which water can evaporate. Plants growing in dry or windy places have smaller leaves than those growing in more sheltered places. The Heather has very small leaves with their edges rolled under to shelter the pores from the wind (Fig. 49).

Plants living in salt marshes or in the desert can get very little water. Their leaves are often very small. In the Cactus the leaves are only small spines, and the stem is very thick and fleshy. It serves as a storehouse for water. The sea-side plant Mesembryanthemum has very fleshy stems and leaves.

During the winter the roots of plants are affected by the cold and cannot take in much water. So there is the danger that more water will be lost by the leaves than can be replaced from the roots. Plants get over this difficulty in different ways. Annuals die in the autumn leaving only their seeds to survive the winter. Other plants have leaves with very thick coverings and few pores, and so they give out little water. These plants are able to keep their leaves throughout the winter and so are called ***evergreens***. Most trees in this country shed all their leaves at the beginning of winter and are said to be ***deciduous***, but some are evergreen like the Holly.

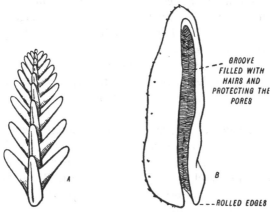

Fig. 49. *A*, Branch of Heather seen under a lens, showing small leaves. *B*, A single leaf showing groove, closely filled with hairs.

Leaf fall

Before a leaf falls in the autumn, cork begins to form where the leaf stalk is joined to the stem. Sap can no longer pass from the twig into the leaf, so it is no longer nourished. The leaf hangs on for a time by the veins which are stronger than the rest of the stalk as they contain wood. This part snaps when the leaf is blown by the wind.

Photosynthesis

Plants obtain their food from two sources. We know that the root hairs of plants take in water and salts from the soil.

The chemical compounds which make up protoplasm contain water, carbon, nitrogen, sulphur, phosphorus, potassium, calcium, magnesium, iron and other substances. With the exception of carbon all these substances are taken in by the root hairs. Nitrogen is taken in in the form of nitrates.

Carbon is obtained from the carbon dioxide gas which goes into the leaves from the air when the pores are open. The carbon dioxide gas dissolves in the water in the leaf and passes into the cells of the leaf. In the cells the chlorophyll causes the carbon dioxide to join chemically with the water to form sugar. Oxygen is given off and is passed out through the pores. This process is called *photosynthesis*.

The sugar formed in the leaves is quickly turned into starch. This will not dissolve in water or in the sap so it forms grains of starch (Expt. 12, Chap. 4). In some plants the starch is stored in other parts of the plant, such as in the swollen underground stems of the Potato. Before the starch can go away from the leaves it is changed again into sugar which will dissolve in the sap. Then it is carried in solution to the stem where it is changed once more into starch. A very simple experiment can be done to show that leaves, or other parts of a plant, contain starch (Expts. 13 and 14, Chap. 4).

Photosynthesis cannot take place unless there is chlorophyll in the leaf. The leaves of some Privet and Laurel plants are yellow, or have patches of yellow on the green leaves. Take such a leaf and draw it, showing the yellow and green patches. Then test the leaf for starch. You will find that the blackish-blue colour formed by iodine when starch is present appears only where the green colour had been.

Animals depend on plants

Animals depend on plants for two things:

(1) Throughout their lives animals are taking in oxygen from the air and breathing out carbon dioxide. Unless more oxygen were put into the air, it would soon all be used up. Plants during

photosynthesis give off oxygen which is passed through the pores into the air. Plants thus give animals their oxygen.

(2) Animals also depend on plants for their food. Plants feed on salts of different kinds, on water and on carbon dioxide. From these they can build up the compound substances which the animals require. Animals obtain these chemical compounds by eating plants, or by eating animals that have eaten plants. For instance, cats eat birds which have probably eaten the seeds of plants. Large fishes eat smaller fishes. These may live on smaller animals in the water which in turn live on the water plants. Spiders eat flies. Flies eat sugar which has been made in plants. Without green plants animals would get no food, so all animal life depends on plant life.

Respiration

Plants, like animals, have to breathe or **respire** if they are to live. That is, they must take in oxygen and give out carbon dioxide. The oxygen gas, which is part of the air, enters the leaves by the pores. Stems have breathing holes (Fig. 32) to let in the oxygen. Roots get their oxygen from the water in the soil. In a well-drained soil there is plenty of air and so plenty of oxygen. In a water-logged soil (Chap. 6) no oxygen can get down from the air above, the roots cannot get their oxygen, so they die.

Water plants obtain their oxygen from the water. The oxygen is dissolved in the water which passes through the skin of the plant. The leaves are often very much split up to give a larger surface for absorbing the oxygen dissolved in the water (Fig. 43).

We learned that during photosynthesis carbon dioxide goes into the leaves through the pores, and oxygen passes out. Now we know that when a plant respires oxygen is taken in and carbon dioxide is passed out. When it is dark, or when plants are put into a dark place, respiration alone occurs. Photosynthesis can only go on when it is light. In the light, although respiration continues as usual, photosynthesis goes on more quickly so that carbon dioxide gas goes in and oxygen comes out of the leaf.

So it seems as if plants take in carbon dioxide and give out oxygen in the light, and do the reverse when it is dark.

Chapter 5

FLOWERLESS PLANTS

All plants can be put into one of two groups: (1) Flowering Plants, (2) Flowerless Plants.

The flowers of some plants are very obvious because they are brightly coloured. Other plants, such as the Oak and Elm, have flowers that are not brightly coloured, so they are not easily seen.

There are many plants that do not have flowers at all.

In this chapter we shall study a few of these Flowerless Plants.

Yeast

Yeast is a very peculiar, little, colourless plant. Like the Amoeba (p. 62) it consists of one cell (Expt. 1, Chap. 5). When you look at the Yeast cells under a microscope you will see that they are oval in shape. Most of the cells have one or more little bumps on them (Fig. 50). A Yeast cell reproduces by this very simple way of forming a

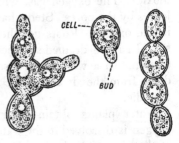

Fig. 50. Yeast cells under the microscope. Successive stages of budding are seen, leading to the formation of chains of cells.

new cell, or **bud**, on its side. Later on the bud becomes free. Sometimes a bud, before it comes off, forms another bud on itself. A whole chain of Yeast cells may be formed in this way (Fig. 50). Yeast grows very rapidly in a liquid containing sugar as well as other substances.

The uses of Yeast

Yeast is used in brewing, in making wine and for baking bread. Brewers' yeast and bakers' yeast are cultivated, but wine yeast is found in the soil in vineyards, and also forms a "bloom" on the Grapes.

Fermentation

Yeast is very important because of its power to *ferment* sugar. When the sugar is fermented it is split up into alcohol and carbon dioxide. Only 2 per cent. of the sugar is used as food, helping to build up the substance of.the Yeast plant, the rest is fermented. A very small amount of yeast will therefore give a large amount of alcohol.

When wine is made, the Grapes are first pressed. The bloom on the fruit, which is really yeast, at once begins fermenting the sugar contained in the juice. If the crushed purple Grapes are left with the juice during fermentation, a red wine is produced; if the grape-skins are removed before fermentation, a white wine results, from purple or from yellow-green Grapes. Dry wines are those in which all the sugar has been changed into alcohol, in sweet wines some sugar remains unfermented. For sparkling wines, the last stages of fermentation go on in the bottles. The carbon dioxide is then retained in the bottles, setting up a considerable pressure. When a bottle of sparkling wine is uncorked, bubbles of carbon dioxide can be seen rising to the top of the liquid. After the carbon dioxide has been given off, the wine is said to be "flat".

Beer is made from Barley grains. The grains are wetted and are spread a few inches deep on the floor. They are kept warm and from time to time they are turned and moistened. The grains begin to sprout and the starch in them is changed into sugar. The germinated grains are then killed by heating them, so *malt* is formed. The malt is ground and washed with hot water. The liquid is filtered off and is boiled with Hops to give it a certain flavour. Yeast is finally added to this liquid when it is cool. The yeast increases six to eight times its own weight by division of its cells. It ferments the sugar in the liquid and beer results.

When bakers put yeast into dough, the sugar in the flour is fermented. The carbon dioxide gas which is formed by the fermenting sugar makes holes in the dough and so makes the bread light (Expt. 2, Chap. 5). The alcohol formed disappears in the baking.

Bacteria

Bacteria, or "germs", or "microbes", as they are often called, are really colourless plants. The plant consists of one cell, which is very much smaller than that of the Yeast. Some Bacteria are only about $\frac{1}{25000}$ in. long. Although very small the different kinds of Bacteria vary very much in shape (Fig. 51).

Bacteria reproduce by splitting into two as the Amoeba does. Under good conditions they reproduce very quickly. They may divide as often as once every twenty minutes.

Occasionally a Bacterium forms a thick wall round itself and is then called a **spore**. Heat kills ordinary Bacteria. Spores, however, can only be killed if they are heated to a temperature of over 120° C. Spores can be blown about from place to place by the wind. Those Bacteria which do not form spores are carried about in the water attached to the dust particles in the air.

Bacteria are very important for several reasons. When animals or plants die we say that their bodies become rotten. What happens is that certain Bacteria are blown on to them by the wind. The Bacteria live and feed on the dead bodies. They break down the compound substances of which plants and animals are made into simple ones, and so the dead remains are got rid of. Evil-smelling gases are formed which are blown away. Among the substances left are compounds of nitrogen, which are turned into nitrates by other Bacteria.

Green plants need nitrogen in the form of nitrates for food. This they use to build up **proteins** (see p. 112), which in turn will be eaten by animals. Green plants cannot use the nitrogen that is present in the air. They can only take it in as nitrates from the soil. The nitrates formed by Bacteria from the dead bodies of plants and animals are food for green plants.

Certain Bacteria use nitrogen gas from the air to build up their protoplasm. When they die they rot like other plants, and the nitrogen again appears as nitrates. Many of these Bacteria live

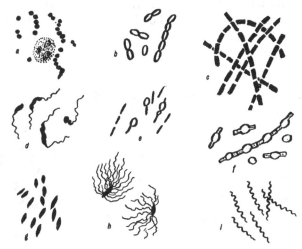

Fig. 51. Bacteria. Various kinds are shown. Some are in chains.
Spores are forming in *e* and *f*.

in the soil. Some of them live in peculiar, little swellings which are found on the roots of all plants belonging to the Pea family.

A farmer can enrich his land with nitrates by growing plants of the Pea family. If the roots are ploughed into the soil, nitrogen compounds are added because the Bacteria in the roots have taken into themselves nitrogen from the air.

Some Bacteria live and feed on living animals, not on dead bodies. They are very harmful. These Bacteria can get into our bodies in two ways. Firstly, they may enter through a wound if it is not kept clean. Secondly, they may enter through our mouths either by getting germs on our fingers and then putting them into our mouths, or by breathing the Bacteria from the air after a person has coughed or sneezed. When inside our bodies the Bacteria give out waste materials which cause diseases. We shall learn more about these diseases in Chapter 11.

Bacteria often cause food to go bad or putrefy. When food is to be preserved in tins or bottles it is first cooked. The heat kills all the Bacteria. The tins and bottles are then made airtight so that no Bacteria can get in. Food is often kept in cold storage. Cold does not kill the Bacteria but it does stop their activities and reproduction. So the food does not go bad.

Fungi

If you leave any food in a damp atmosphere exposed to the air, it becomes mouldy. Moulds, like Yeast and Bacteria, are colour-

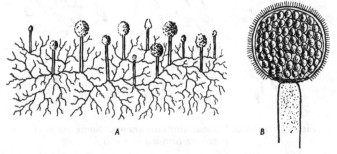

Fig. 52. Mucor. *A* shows the branching threads with sporangia at the tips of the upright branches. *B* is a sporangium more magnified.

less plants. The most common Mould is the white fluffy Mould which quickly grows on food that is left exposed to the air.

Place a piece of damp bread on a saucer under a bell-jar and leave it for several days. After some time you will see this white, fluffy Mould on the bread, with a number of small, black spots on it. This Mould is called the Pin-spot Mould or Mucor. Look at it with a lens (Expt. 4, Chap. 5). You will see a number of thin threads. Sometimes branches of the threads grow upwards.

On the top of each branch is a knob which looks like a pin-head. Inside these knobs, or spore boxes, or *sporangia* as they are called, are numerous spores (Fig. 52). The wall of the knob bursts when the spores are ripe, and the spores are shot out. These spores float about in the air and get on to food. So all food when damp, if exposed to the air, will go mouldy.

You will not find Mould growing far down into the food because the threads must have oxygen. For example, if Mould gets into a jar of jam, the top layer will be mouldy, but the bottom layer of jam will be free from Mould.

Mucor belongs to the group of colourless plants called Fungi. By colourless plants, we do not mean that the plants are without colour. Colourless plants are those that do not contain chlorophyll. One very common Mould which soon grows on food left exposed to the air is the Blue-green Mould. The green colour is not chlorophyll, so it is called a colourless plant. The green streaks in gorgonzola cheese are made by this Mould which is put into cheese.

The Mushroom

The Mushroom is another fungus (Fig. 53). The plant consists chiefly of threads similar to those of Mucor. These are not usually seen, as they live in the ground (Expt. 5, Chap. 5). As the plant has no chlorophyll it cannot make its own food. Mushrooms live on decaying plant or animal matter in the soil.

Like the Moulds, the Mushroom reproduces by means of spores. The part of the plant which we eat contains the spores. They grow on the *gills* on the underside of the head (Fig. 54, Expts. 6 and 7, Chap. 5).

Other Fungi

The Fungi are a very large group of plants. A good collection of Fungi can be made during the autumn (Expt. 8, Chap. 5). The reference books mentioned in Appendix C will enable you to recognise any Fungi that you may find. The bulletin issued by the Ministry of Agriculture and Fisheries will tell you which of these Fungi are edible and which are poisonous.

"Fairy Rings" are rather interesting. They are really rings of "toadstools". These Fungi grow in ever-widening rings as they use up the food in the soil within the ring.

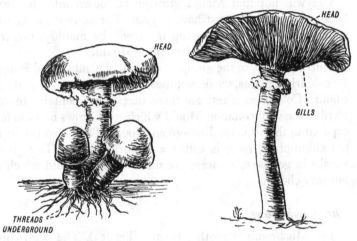

Fig. 53. Mushrooms.

Fig. 54. A Toadstool showing the gills.

A large number of Fungi are very harmful. Some attack man and animals and cause diseases. Ring-worm is a Fungus growing as a "fairy ring" on a person's skin. One form of eczema, Barber's Itch, and Thrush, are also caused by Fungi. Thrush appears as small white patches in the mouths of children. Many Fishes, for example Goldfish and Sticklebacks, are attacked by Fungi and may even be killed.

Fungi also destroy crops such as Wheat, Oats, Fruit and Potatoes. The familiar brown rot on Apples, and the scabs on Potatoes are due to these Fungi.

Other Fungi rot wood. It is very dangerous if the wooden

beams in large buildings are attacked by these Fungi, as the buildings may collapse.

Some lower green plants

So far we have studied flowering plants which have leaves, stems and roots, and colourless plants having a very much simpler structure. There are many simpler green plants.

Algae

The green coating seen on damp tree trunks and palings, the green scum on ponds, and Seaweeds belong to this group of simple green plants.

Most of the plants found in the sea are Seaweeds. Although some Seaweeds are brown or red, they are "green plants". The green colour of their chlorophyll is hidden by the red or brown colour. The latter colours will dissolve in fresh water, then the green colour can be seen.

Mosses

Mosses are simple plants found in most parts of the world (Expt. 9, Chap. 5). Look at a plant of the Moss most commonly found on the ground or on walls (Fig. 55). It consists of root-like structures, stems and leaves. You will often see small capsules at the end of long stalks. These capsules contain the spores. When young they are green, but finally they turn brown. When the spores are ripe, the lid of the capsule comes off. Underneath the lid there is a ring of teeth, which change their shape in wet and dry weather. When it is wet they close down but in dry weather they open out. The spores are blown away by the wind and so scattered. Later the spores grow into new Moss plants.

Fig. 55. A Moss plant.

Ferns

Fern plants are more nearly like flowering plants than any other plants yet described in this chapter (Expt. 10, Chap. 5).

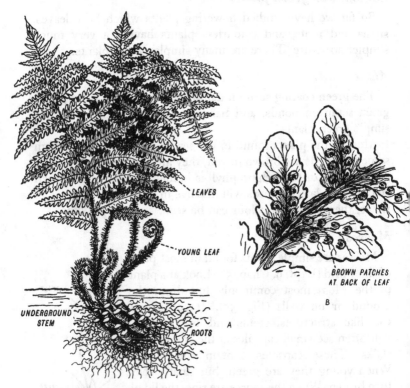

Fig. 56. *A*, A Fern plant. *B*, Part of a leaf highly magnified to show brown patches at the back of the leaf.

If you look at the so-called Male Fern you will see that it consists of an underground stem from which roots and leaves grow (Fig. 56 *A*). As in flowering plants the roots take in salts in solution from the soil. Photosynthesis and transpiration also go on in the leaves.

During the winter the Fern leaves die down and new ones grow the following spring.

If you look at the back of a Fern leaf during the autumn you will see that it is covered with a number of brown patches (Fig. 56 B). Each brown patch is made by a brown cover. Remove one of these covers and look underneath it with a lens. You will see a number of short stalks with a knob at the end of each. These are sporangia in which spores are formed. When the spores are ripe, and the weather is dry, these knobs burst and the spores are shot out. The spores eventually grow into new Fern plants.

Chapter 6

THE SOIL

The formation of soil

Soil is made up of two things, powdered rock and decayed animal and vegetable matter. The rocks are powdered by the action of water, frost, heat, wind, air and plants (Fig. 57).

Rain assists in breaking up the soil by dissolving the soluble substances which are in the ground. For instance, the limestone rock called marl is made of carbonate-of-lime and clay. As the rain falls through the air it dissolves some of the carbon dioxide gas contained in the air, forming a weak acid. The carbonate-of-lime easily dissolves in this acid, leaving behind the clay.

Running water, such as that of streams and rivers, wears away the rocks of the river bed or those along the banks. This material is then carried away and deposited elsewhere. The sea, of course, also wears away the rocks round the coast.

Frost also takes a part in breaking up the rocks. During the cold weather, water may collect in the cracks and holes in the rocks. The water then freezes, and as it expands whilst freezing, it makes the cracks larger. Farmers often plough in the autumn, especially if they have clay soil in their fields. (See p. 59.) Rain then falls on to the soil. If this water freezes, then the soil may be broken up.

The **heat** of the sun causes rocks at the surface to expand. During the night time when the rocks become cool, they contract. This causes them to split and crumble.

Fig. 57. Formation of soil.

Wind does not help to wear away the rocks, but in places where the soil is not covered with vegetation the wind may blow the soil for some distance. The sand storms of the deserts are clouds of sand, which are blown along by the strong winds.

Plants may help to split up rocks. Those plants which grow on the rocks force their roots down through the cracks, making openings into which air and water can pass.

Animals do not help to split up the rocks, but they help in the formation of soil. Animals living in the soil act as underground ploughs, constantly turning over the soil. We shall learn in Chapter 7 how useful Worms are in the soil. Plants will not grow in soil containing no animal or vegetable matter. Animal manure, the dead bodies of animals, and decaying plants all mix with the soil.

Different types of soil

Put a small amount of soil into a jar containing water. Stir this up well and then allow it to settle. Look very carefully at the jar. You will see that the coarse, heavy particles are at the bottom. These are either stones, gravel or **sand** particles. On top of this there are the very fine particles of **clay**. Floating on the top of the water is the animal or vegetable matter which is called **humus** (Fig. 58).

Soils are classified according to the amount of sand, clay or humus which they contain. A soil containing a normal amount of each is called a **loam**. As the amount of sand increases, the soil is called a **sandy loam**, or **sand** if there is a large proportion of sand. Similarly as the amount of clay increases in proportion to the sand, the soil is called a **clay loam** or **clay**. If the soil contains much humus it is called a **peaty** soil.

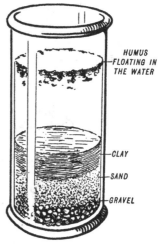

Fig. 58. Experiment to show the constituents of soil.

After it has been raining, some soils seem to dry very quickly whilst others remain wet. All soils are porous, as there are small spaces between the particles of soil. This space in the soil is called the **pore space**. Experiment 5 (p. 180) will show you that the smaller the particles the greater the pore space. In coarse, sandy soils the pore space is about 25 per cent. of the whole soil, but in a clay soil it may be 50 per cent. This pore space should be partly filled with air if plants are to thrive, as roots cannot grow without oxygen. Around each particle of soil there is a film of water. The smaller the particles of soil, the greater the number of films. So there is more water in a fine soil than there is in a coarse soil.

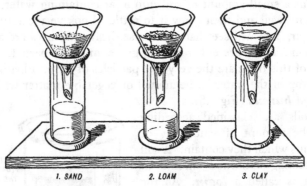

Fig. 59. Experiment to show the water-holding capacity of different soils.

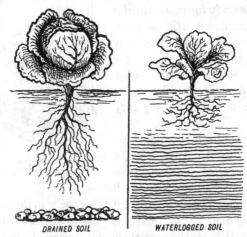

DRAINED SOIL   WATERLOGGED SOIL

Fig. 60. Diagram showing the effect of a water-logged soil on the growth of a plant.

The following experiment shows what kind of soil is able to hold the most water.

Take three filter funnels and put a filter paper in each. Then rest each one on the top of a gas jar (Fig. 59). Put dried sand into the first, dried loam into the second, and powdered, dried clay into the third, taking care that the same amount of each is used. Carefully pour equal quantities of water on to each and see how long it takes for the water to run through the soil.

Sand allows water to run through very quickly, but the water passes very slowly through the clay. A sandy soil is not good for plants because the water runs through too quickly. The surface soil soon becomes dry, so plants would not have sufficient moisture. A clay soil is also bad for most plants. As the water can only run through slowly, the pore spaces may be filled with water for a considerable time. No oxygen could then get to the roots and the plants would not thrive (Fig. 60). If the pore space is almost continuously filled with water, as it often is in flat meadows along the banks of streams, the soil is said to be *water-logged*. A mixture of clay and sand together with humus is the best kind of soil for most plants.

Flocculation

When clay is wet, the small particles stick together, preventing any water from passing to the soil beneath. In some way farmers have to make this soil useful for plants. They often dig ashes, sand or lime into the soil. The previous experiment tells us why ashes and sand are used. Another simple experiment will show us what action the lime has on the soil.

Thoroughly mix up some clay and water and pour some of the liquid into two glass jars. Into one of the jars put some powdered lime and leave it for a short time. You will see that the water in the jar containing lime is the first to become clear (Fig. 61). The lime causes the clay particles to stick together forming larger particles. The clay is then said to be *flocculated*. Water can pass through flocculated clay.

If you put your hand on sand on a very hot day, the sand will be so hot that it will almost burn you. If you put your hand on clay you will feel that it is not so hot. Sand gets hot very quickly, but it also cools quickly. A well-drained soil, similarly, gets warm fairly quickly. Clay soil, on the other hand, is usually wet and so it requires much heat to raise its temperature. Plants do not grow well in cold soil. A farmer may improve his clay soil by draining it in some way, as well as by mixing ashes, sand or lime with it. Clay soil is slow to warm in the spring time but it cools slowly after the heat of summer has gone.

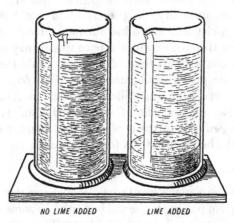

NO LIME ADDED   LIME ADDED

Fig. 61. Diagram to show the action of lime on clay.

Plants in clay soil are later in growing than those in sandy soil. The latter kind of soil is an "early" soil, whereas the former kind is a "late" soil.

Most soils contain the necessary amounts of salts for plants. Where the plants are growing wild, the salts removed from the soil by the plants are returned to it when they decay. All plants do not use up the same salts from the soil. Some plants require more of certain salts than others, for instance, Peas and Beans require different salts from Cabbages. If we grew the same kind of plants on a piece of ground for several years, the soil would soon be without the particular salts that the plants were using. These salts could be added to the soil by putting manure on the ground. A much better way is to grow a crop one year that requires different salts from those used by the crop grown the previous year. In the third year, a crop using yet other salts could be grown. Meanwhile the soil would gradually become richer in the salts which had been used up in previous years. This changing of crops is called the *rotation of crops*.

Effect of soil conditions on plant growth

We have already learned that the plant is affected by the condition of the soil. The type of plants growing in any place, and the health of the plants, depend on the following things:

1. The type of soil.
2. The amount of water in the soil.
3. The temperature of the soil.
4. The amount and kind of food present in the soil.

Chapter 7

SOME SIMPLE ANIMALS

All animals can be placed in one of two large groups:

1. Animals with backbones: these are constructed more or less like ourselves.

2. Animals without backbones: these are quite unlike us.

By considering just one thing, we can tell into which of these two groups any animal should be put. Every animal with a backbone has a skeleton consisting of bones, which make the body rigid (see Chapter 10). Backboneless animals, on the other hand, have no bones, although some have a hard outer covering which makes the body rigid, for example, the shell of a Beetle or of a Crayfish (Fig. 69).

The Amoeba

The Amoeba (Fig. 62) is a very tiny animal, about one-fiftieth of an inch across, which is found in the mud at the bottom of ponds. (If you have a microscope in your school, your teacher will show you an Amoeba, as you cannot see it with your naked eye.) It looks like a small piece of soft colourless jelly which is always changing its shape. It is not divided up into cells.

MOVEMENT. When the Amoeba moves, its soft jelly-like substance, which is protoplasm, bulges out on one side, then the remainder flows after it in the same direction.

FEEDING. If an Amoeba meets a plant smaller than itself, the protoplasm of the Amoeba flows round the plant. So the Amoeba's food becomes surrounded by a drop of water inside the animal (Fig. 62). The plant is then digested, that is, changed into a substance which will dissolve in water (see Chapter 10). This is done by a solution which passes into the drop of water from the protoplasm. The digested food next passes from the drop into the protoplasm. The animal then flows away from that part of the food which was not digested and leaves it behind.

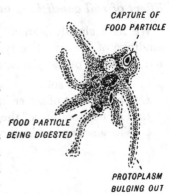

Fig. 62. Amoeba.

BREATHING. We cannot see the Amoeba breathing, but oxygen, dissolved in the water, passes into its body, and carbon dioxide passes out again, over the whole surface of the animal.

REPRODUCTION. When an Amoeba is fully grown, it divides into two (Fig. 62). Two daughter Amoebas are thus formed, and there is no parent left to die of old age. When larger animals than Amoeba are growing, the cells of which they are made divide in this same way, but the daughter cells remain together instead of separating from one another.

Although the Amoeba consists of only one cell, we have learned that this cell can move, feed, breathe, get rid of waste matter, reproduce. It will even move away if a drop of acid is put into the water.

SIMILAR ANIMALS CAUSING DISEASE. You have probably heard of people having sleeping sickness, malaria fever, or dysentery, which is a very bad form of diarrhoea. These diseases are caused by small Amoeba-like animals in the blood. In Chapter 10 you will learn how these animals get into our blood.

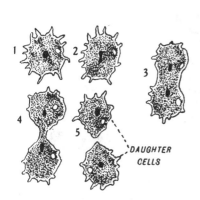

Fig. 63. Amoeba dividing.

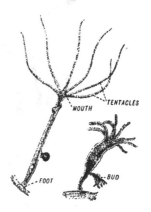

Fig. 64. Complete Hydra. The animal on the left is fully extended, that on the right is half extended.

The Hydra

The Hydra is a small animal, usually about one-quarter of an inch in length, although it can stretch itself to more than twice that size, or shrink to a small lump (Fig. 64). Hydras are either brown or green in colour, and can be found fixed to the weeds in ponds. Next time you go fishing, try to find some. They are very easy to see in a jar in school, but difficult to see in a pond, so you must bring some weeds to school. If you leave the weed in a jar for a day without disturbing it, you may see some Hydras. These animals soon die if not properly cared for (see Appendix D).

If you live near the sea you can easily look at the Sea-anemone, instead of Hydra. It is very much like an enormous Hydra.

A Hydra is usually attached to a weed at one end, called the *foot*, whilst at the other end there is a **mouth**, which leads into a space inside called the **stomach**. Round the mouth are a

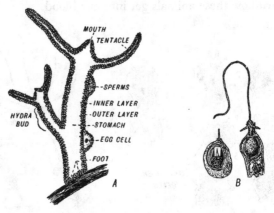

Fig. 65. *A*, Section of Hydra. *B*, On the left the thread cell is coiled inside the cell: on the right the thread has been shot out.

number of fine threads, the **tentacles** (Fig. 65 *A*). These move about in the water, feeling if anything is near. If anything touches a tentacle, or if the water is disturbed, it immediately shortens up to the body.

MOVEMENT. A Hydra may stay in the same place for a long time. When it moves it slowly glides along on its foot.

FEEDING. On the tentacles are a number of **thread cells** (Fig. 65 *B*). These can only be seen through a microscope. When a tiny animal touches the "trigger" of a thread cell, out shoots the thread, which is sticky and poisonous. Then the tentacles close over the numbed prey, which is drawn into the mouth, and then passed into the stomach, where it is digested. The stomach is really like a bag, surrounded by two layers of cells, and has only one opening which is the mouth (Fig. 65). Undigested food, therefore, has to be passed out through the mouth again.

BREATHING. The animal takes in oxygen, which is in solution in the water, and gives out carbon dioxide, over the whole surface of its body.

REPRODUCTION. If you can keep Hydra in your science room, you will find that they often have lumps on their bodies, which can easily be seen with the naked eye. The larger lumps are **buds**. At the end of each bud a mouth and tentacles are formed. A wall then grows across the base, cutting the bud off from the parent. Then the young Hydra moves away and starts life by itself. A Hydra is very different from an Amoeba, because when it reproduces a parent is left behind, which will live for some time and later on will die, while its offspring continues alive.

The Hydra can reproduce in another way. You may see smaller lumps on the animal. In those nearer the foot, one large cell grows. This is an **egg**, although it looks very different

Fig. 66. Single sperm. Magnified 4000-fold.

from a bird's egg (Fig. 65). In the lumps nearer the mouth, there are a larger number of tiny cells, which have a very strange shape (Fig. 66). They have a "head" and a long "tail". These strange cells are called **sperms**, and when the lump in which they are formed bursts, they swim vigorously in the water by means of their tails. If a sperm touches an egg, it joins with it, and then the egg can grow into a new Hydra. We say that the sperms **fertilise** the eggs. Only one sperm can fertilise one egg.

If part of a Hydra is cut off, the remaining part can grow again into a complete animal.

Corals

You can often see a Hydra with several buds on it. Imagine that, instead of becoming free, each bud grows to a full-sized Hydra, but remains attached to the parent. Then imagine that each of the buds, buds again, and so on until you have a small "colony" of animals, which seem to be branching almost like a tree (Fig. 2). Each animal of a coral colony makes or secretes a hard shell underneath itself of calcium carbonate (which is something like limestone); then of course it cannot move. After a time the animals die, leaving their shells, which are the Coral as we know it. Pink Coral is the most familiar kind, because we often see pink coral necklaces. The Great Barrier Reef of Australia, which is 1000 miles long and 50 miles wide, has been made by these tiny animals.

The Earthworm

Go into your garden and look for a worm. If it has been raining you will see worms on top of the ground. If it has been very dry, however, you will have to dig far down into the soil before you find one, because worms die if they are dry, and so remain far down in their burrows until the drought has passed.

Now look at your worm closely. Its body is made up of a number of rings or **segments** joined together. Let your worm move along, and you will see that it usually pushes its more pointed end forward. This is the **head**, which has neither ears nor eyes like we have, but only a **mouth**. If you hold your worm and look between the first two segments through a lens, you will see the mouth. On the last segment there is another opening, the **anus**, through which the worm gets rid of waste matter (Fig. 67).

MOVEMENT. On the sides of every segment are **bristles**, which you can feel if you rub your fingers along the worm. These give the worm a foothold as it moves along. It moves by pushing its head end forward, then drawing its tail up near to its head.

BREATHING. The worm breathes through its skin. The passing in of oxygen and passing out of carbon dioxide gas can only take place through living cells. The skin is kept damp by being covered with a slimy moist substance. This is important, because if the cells dried up they would die, and then the worm could not breathe.

FEEDING. Worms feed by eating the soil. The soil is passed into a "stomach", where the microbes which are in it, and which are the real food of the worm, are digested. The results of the digestion are then passed into the **blood**. The digested food is carried to all parts of the body by the blood, which flows in veins. The blood is coloured red like our own, and, besides carrying food, it carries oxygen and carbon dioxide from the skin to and from all parts of the body.

Undigested food is not passed out through the mouth again, as in the case of the Hydra, but passes from the "stomach" along a tube until it reaches the anus, where it passes out. This is so in all animals more complicated in structure than the Hydra. The waste matter of the worm forms the coiled pieces of soil, often seen on lawns. These are called **worm casts**.

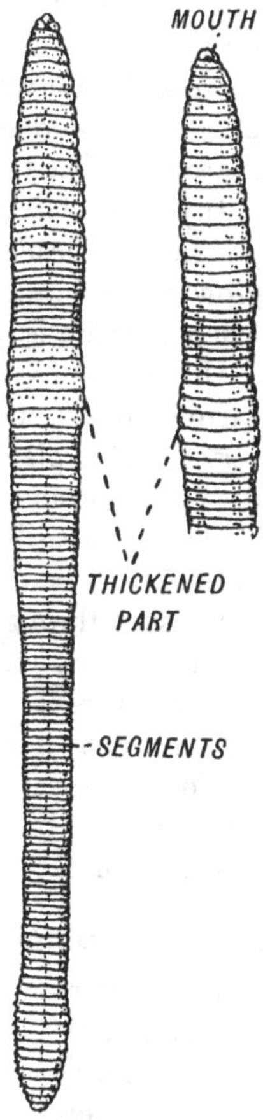

SENSES. A worm cannot see or hear, but it has **nerves** from its skin to its brain. With these it can feel the ground shake when an enemy is near, just as you can feel a heavy lorry shaking the ground as it passes by.

Fig. 67. Earthworm. Left figure from upper side; right figure from the lower side.

5-2

REPRODUCTION. A little way behind the head you will see a thickened part in all large worms. Round this part a case is formed from time to time, into which the worm passes eggs and sperms; then it wriggles backwards out of the case. The ends of the case, or *cocoon* (as it is called), close up and, inside it, the sperms fertilise the eggs. These then grow into new worms.

If you accidentally cut a worm in two when gardening, it does not die. The head end can grow a new tail, but the tail end cannot grow a new head, so it dies. Do you remember what happens when a Hydra is cut in two?

USES OF WORMS. Worms make burrows in the soil, which are narrow tunnels ending in small round chambers. The worms are very useful to plants for a number of reasons. Air can pass down their burrows and so give oxygen to plant roots which always need it. Water can easily drain away down these spaces, and so prevent the soil from being waterlogged (see Chapter 6). The soil which has passed through a worm's body is finely ground up, so the tiny root hairs, which were described previously, can easily push their way through the soil. The worms also drag leaves into their burrows, thus manuring the soil.

Worms are very plentiful in soil, unless it is sandy and therefore too dry, but many are eaten by other animals, such as thrushes, blackbirds, toads, frogs, lizards.

Animals similar to the Earthworm

There are many different kinds of worms, living in all sorts of. places, on land, in the water, or even inside other animals.

The Horse Leech is an example of a worm living in water, in ponds, streams or canals (Fig. 68). At each end of its body it has a sucker, by which it clings to any object. The mouth is in the middle of the front sucker. The Leech fastens itself to an animal with this sucker, makes a small wound with its jaws, then sucks the blood, so gradually killing its prey.

Leeches something like the Horse Leech were at one time, and are occasionally nowadays, used by doctors to "bleed" patients suffering from certain diseases. In tropical forests leeches may attack men.

Some kinds of worms are often found inside larger animals, such as horses, dogs, etc., and are even found in human beings. The most common of these worms is the Threadworm, a small, white worm about half an inch long, very frequently found in children.

Fig. 68. The common Horse Leech.

Shrimps, Lobsters and Crabs

These animals, together with Prawns and Crayfishes, all belong to one group. You have seen most of them in a fish shop dead, but probably few of you have seen any of them alive. All these animals, except the Crayfish (Fig. 69), live in the sea, but the Crayfish lives in streams. There you will also find Fresh-water Shrimps, which are not quite like the Shrimps from the sea which you buy in shops.

All these animals have a hard shell covering them, to protect their bodies from other animals. Their colour is only reddish pink after they have been boiled. When they are alive, Lobsters are blue, Crabs are brown or green, while Shrimps are grey. They cannot grow in the same way as we do. Their shell cannot stretch, so from time to time it splits along the back, and the animal wriggles out of it. Then the animal hides for several days, because its enemies could easily kill and eat its soft body. Meanwhile it grows until its new shell, which has formed, hardens.

Most of these animals have eyes which are on stalks. They also have a large pair of claws for catching their prey, and four

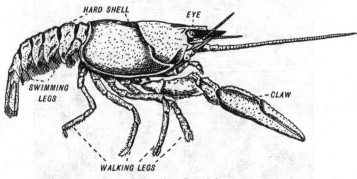

Fig. 69. Crayfish.

pairs of legs on which they walk (Crabs walk sideways). In addition, they have many other smaller "legs" which Shrimps and Prawns use for swimming while female Lobsters and Crayfishes carry their eggs stuck to them.

The Woodlouse (no connection with the Louse) is a relative of Shrimps and Crabs which is often found in houses and gardens. It lives on land, not in the water like its relatives. One kind can roll up to defend itself from enemies.

Unlike the Earthworm, all these animals die if cut into two. But if they should lose a claw or a leg, a new one can grow.

Spiders

All of you, I am sure, have seen a Spider, yet do you know how many legs it has? Insects, as you will learn later, have only six legs, whereas a Spider has eight, so a Spider is not an insect. At the end of each leg there is a jointed hook, by means of which the Spider can run along the finest thread (Fig. 70).

Its body is divided into two parts, a small head and chest joined together, and a fatter, hinder part called the **abdomen** (Fig. 70).

The Spider has eight small eyes, yet it cannot see very well. It has, on its head, some strange jaws, which are like claws, and feelers. Poison passes down tubes into these jaws and out at their tips. A Spider wounds its prey with its jaws, and at the same time poisons it.

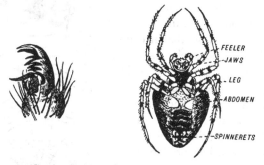

FEELER
JAWS
LEG
ABDOMEN
SPINNERETS

Fig. 70. Garden Spider. On the left, spider's foot enlarged.

THE WEB. One of the most interesting things about Spiders is the way they make their webs, which are really traps for catching their food. The webs are made of fine, silk threads. If you can find a large Spider, look underneath its abdomen with a lens and you will see six small lumps, called **spinnerets**. Out of these comes a gummy liquid, which, as it dries in the air, forms a fine silken thread.

When spinning its web, the Spider first makes a framework of threads which looks like a wheel with spokes. This is fastened at four or five points to leaves or branches. If any thread is too loose, the Spider pulls it up with her claws. After this the spiral part of the web is woven, but now a sticky thread is used.

Some Spiders remain in the centre of their web until an insect is caught in it. Others, like the Garden Spider (which has a cross on its back), hide a little way away, with one foot touching a silken thread which is fastened to the web. This is a "telegraph wire", for when an insect flies into the web, the web shakes and the thread is pulled. The Spider feels it and knows that her next meal is ready. The Spider often kills insects bigger than herself. She leaves her prey hanging by two threads, and turns it round and round, weaving a web around it so that it cannot move. Then she sucks its blood, or carries it into her "larder" (Fig. 71). Some Spiders do not make their webs in the form of a spiral, but weave numerous threads at random in all directions.

There are other Spiders which make no webs but hunt their prey.

REPRODUCTION. Male Spiders do not make webs of their own, but steal food from the female's web. The females are often larger than the males, and sometimes eat their

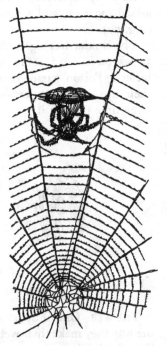

Fig. 71. A Spider wrapping silk round a victim.

husbands, although the latter may perform strange dances to gain their favour.

Females find a sheltered spot, and weave a silk case or cocoon, which is sometimes $\frac{3}{4}$ inch across, within which they lay about one hundred small, yellow eggs. These are laid in September. They may hatch but before the winter or not until the following spring.

The Water Snail

The Water Snail belongs to a group of animals most of which have **shells**. These shells may be in one piece as in Snails, Periwinkles, Whelks and Limpets, or in two pieces hinged together, as in the Mussel (Fig. 72) and Oyster. Water Snails are the simplest to study because they can easily be kept in an aquarium (see Appendix D). They are found in ponds.

Fig. 72. Shell of Freshwater Mussel showing the two pieces of the shell.

Put a Water Snail in a glass jar full of water and look at it. Its shell is a spiral of six or seven turns, when fully grown. The oldest part is at the top. The shell protects the rest of the body, which is very soft. Only a large fleshy part called the **foot**, and the **head**, ever come out of the shell (Fig. 73). On the head you will see a pair of **tentacles** or feelers, two **eyes** on small lumps, and a **mouth** on the underside.

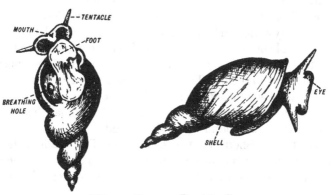

Fig. 73. Common Pond Snail.

MOVEMENT. The Snail glides on its foot; it can even move along upside down on top of the water.

FEEDING. Watch a Snail moving up the side of the aquarium, and you will see its mouth opening and closing, showing an orange speck. This is its **tongue**, which has many rows of hard teeth on it, and is like a file or rasp. The animal rasps away the weeds, or the green "scum" growing on the sides of the tank.

BREATHING. At the edge of the shell, on the right side of the Snail, you will see a hole. When a Snail wants to breathe, it comes to the surface of the water with this hole on top. Air containing carbon dioxide gas is given out, and fresh air, which gives the animal the oxygen it needs, is taken in through the hole into a space or **lung**. In the lung enough air is stored to last for some time while the Snail is under water.

REPRODUCTION. During the summer the Snail lays pieces of jelly on the weeds. These are about $\frac{1}{4}$ inch wide and $\frac{1}{2}$ to 1 inch long, and have twenty to thirty eggs in them. The young Snails hatch out of the eggs in about one month. They are fully grown in about two years, and may live for five years.

SNAILS AS FOOD. Land Snails are often eaten as food. French people breed Snails for this purpose. Relations of the Snail, such as Periwinkles, Whelks, Mussels and Oysters, are also eaten. The Oyster is fixed to one place like the Coral.

Pearl Oysters, chiefly found in the Pacific and Indian Oceans, are valuable because of pearls found inside them, and also for the mother-of-pearl lining the shell, which is used for making buttons and brooches. True pearls are formed by this kind of Oyster to protect itself against irritation due to very small sand grains which sometimes get inside it. The animal then makes (secretes) a pearl round the grain of sand.

SNAILS AND SLUGS ARE PESTS. Slugs are like Snails without a shell. Land Snails and Slugs do a lot of damage in kitchen and flower gardens, and also in fields, by eating young seedlings and potatoes.

Thrushes, frogs, toads, fowls and ducks will eat Slugs and Snails, so helping us to get rid of them. Repeated liming and sooting the ground also helps.

Chapter 8

INSECTS

The insects are a very large group of animals. There are nearly as many kinds of insects as of all other animals in the world.

The body of an insect is segmented, like that of the Earthworm, but in addition it is divided up into three parts. The **head** and the chest, or **thorax**, are separate, not joined as in the Spider. The first one or two segments of the back part of the body, or **abdomen**, are sometimes very narrow, forming a "waist", as in the Wasp. On an insect's head there is a pair of jointed feelers, and two large eyes. On the thorax there are three pairs of legs and usually two pairs of wings (Fig. 75 C). In beetles the first pair of wings are hard, and in flies the second pair are missing. Some insects have no wings, for example the Stick Insect. Unlike Shrimps and Lobsters, insects have no limbs on the abdomen. There may be a sharp tube at the end of the abdomen for placing eggs where they are to hatch. For instance, Ichneumon flies lay their eggs inside caterpillars. Some insects, like the Wasp, have a sting at the end of their abdomen.

All adult insects breathe air, though in their younger stages they may live in water. Although living in water, some young insects nevertheless breathe in oxygen from the air, for example, gnats, whilst others take in oxygen from the water, in which case they have gills, for instance, the Caddis flies and May flies. All insects living on land have special openings on their bodies which lead into tubes. Air passes through the openings into these tubes, which then carry it to all parts of the body. This is a very different state of affairs from our own bodies, where oxygen is distributed by the blood.

Insects can taste and smell, and some can hear, particularly those which make noises. Grasshoppers chirp, not with the throat, but by rubbing their hind legs against their wings, and they have their ears in their abdomens.

The Cockroach

Cockroaches (Fig. 74) like warmth, so are found chiefly in houses, especially old ones, and even in bakeries. They feed on any animal or vegetable matter they can find.

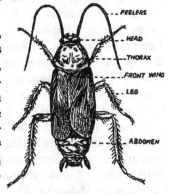

The female has very tiny wings, useless for flight. The female lays eggs during the summer months, sixteen at a time. These are arranged in two rows inside a brown horny case, which the female carries at the end of her abdomen to a sheltered corner where she leaves it. When the young cockroaches hatch out they are like their parents but have no wings. The wings begin to grow

Fig. 74. Cockroach.

after twelve months. The insect is covered with a hard shell like that of the crayfish. This will not stretch, so, as the insect grows, the shell splits from time to time, and the young cockroach with its soft covering wriggles out. The animal swells and then its new shell hardens. The insect loses its skin, or **moults**, seven times in the twelve months that it takes to grow.

Butterflies and Moths

Everyone who has kept silkworms or caterpillars knows that moths and butterflies start life in quite a different way from cockroaches.

One of the most common butterflies throughout the country is the Cabbage White (Fig. 75 C). This butterfly lays its eggs in

May, or in July and August, on the lower side of cabbage and
nasturtium leaves. Caterpillars, or *larvae*, hatch out in seven to
ten days and begin greedily eating the leaves. They moult from
time to time until fully grown, and in colour are pale green with
yellow stripes. The larvae of other butterflies and moths are very

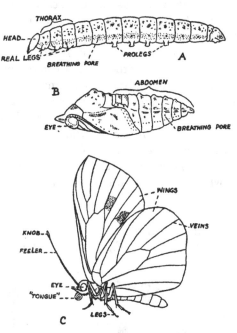

Fig. 75. Life history of the Cabbage Butterfly.
A, Larva. *B*, Pupa. *C*, Adult.

varied in colour and size. The caterpillar has six real legs on the
thorax and several other stump-like legs on the abdomen called
prolegs, with hooks on the tips. With these the caterpillar clings
very firmly as it is climbing about (Fig. 75 *A*). In the middle of
the lower lip of the mouth is a tube through which silk can be
spun.

When fully grown, the caterpillar climbs up to a sheltered spot by making a silken ladder up which it climbs with its prolegs. Then it changes into a chrysalis, or **pupa** (Fig. 75 *B*), which has a very hard covering. The pupa remains still and when touched can only move its abdomen. The caterpillars of some butterflies and moths weave a silken case, called a **cocoon**, round themselves. Inside this they change to a pupa. It is the cocoon of the silkworm which is used for making silk.

If the pupae of the Cabbage White are from eggs laid in May, the adult insects come out of the pupa case in two or three weeks. If they are from the August brood they remain in the pupal stage throughout the next winter. The butterfly has six legs and two pairs of wings strengthened by so-called "veins". Its "tongue" is like a tube, up which it sucks nectar. It is very long and is curled up when not in use (Fig. 75 *C*).

The feelers of moths are generally more hairy than those of butterflies, which usually have a knob at the end of their feelers.

Flies

Flies lay their eggs on animal or vegetable matter which is decaying or "going bad" (Fig. 76 *a*). A female lays up to

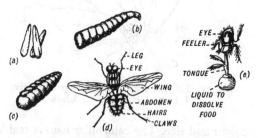

Fig. 76. Stages in life history of Housefly. *a*, Eggs. *b*, Larva. *c*, Pupa. *d*, Adult Fly. *e*, Head of Fly.

150 eggs at a time, and may do so six times. From the eggs the white larvae, called **maggots**, hatch out (Fig. 76 *b*). They feed on decaying matter. Horse manure is a favourite site. A maggot has

no legs, and is pointed at the head end. At the hind end two breathing tubes open. Maggots do not like light.

When fully grown the maggot's skin turns hard and brown, and it changes into a motionless pupa (Fig. 76 c). Later on the adult fly comes out of the pupal case (Fig. 76 d).

In warm weather an egg will grow into a fly in nine to ten days, and a fortnight later the fly will be able to lay eggs. A fly's body is very hairy. It has two wings only. Each leg at its tip has hooks on either side of a pad which is sticky. This enables the fly to walk on the window pane, or on the ceiling. The fly has a "tongue" with which it sucks up liquids. When the fly is about to eat sugar, it first spits out a liquid which dissolves the sugar, and then it sucks up this sugar solution (Fig. 76 e).

Gnats

A Gnat is an English kind of Mosquito. The male gnat can easily be distinguished from the female because he has bushy feelers, and the female has not.

Male gnats feed on the nectar of flowers, but the female stings other animals and draws blood from the living prey. She does this by sticking a pointed tube into the skin of her prey, thus piercing the skin, and then squirting in a liquid which prevents the blood from clotting. Then she sucks up the blood.

The female lays her eggs in ponds, slow-running streams

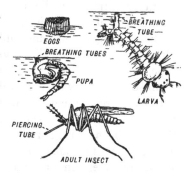

Fig. 77. Life history of a Gnat.

or water butts. She glues them together with a sticky stuff to form a "raft" (Fig. 77), which floats on the surface of the water. The larvae (Fig. 77) hatch out and feed on tiny green plants in the water. From the last segment but one of the larva there grows a tube, which is used for breathing. Although living in the water, the gnat larva gets its oxygen from the air by placing the end of the tube at the surface of the water. When the larva is swimming in the water this tube is closed by a valve. After several weeks the larva turns into a pupa which does not feed, but which can move about in the water. The gnat finally comes out of the pupal case at the surface of the water, and flies away after its wings have hardened.

Insects as disease carriers

Many human diseases are due to germs. Sometimes germs reach our bodies in our food or water, or are blown about in the air, or are carried to us by insects. Many small animals similar to the Amoeba which cause diseases, for instance malaria fever, are also carried from man to man by insects.

The house fly settles on filth, and dirt containing germs sticks to the hairs on its body and legs. These germs may later get on our food. Flies carry the germs of tuberculosis from the spittle on which they alight. It is very important to keep down the number of flies. This can be done by covering up, burning or disinfecting rubbish and dung, so that flies have no breeding places. Killing adult flies by means of fly papers helps a little.

Mosquitoes carry the germs of malaria fever. When a mosquito stings a malaria patient, the insect sucks up germs with the blood as it is feeding. Then, when the mosquito stings another man, there will be malaria germs in the liquid squirted into the wound. The number of mosquitoes can be lessened by pouring oil on the surface of water containing their larvae. Then the larvae cannot push their breathing tubes to the surface to take in fresh air, so they die. English gnats seldom carry diseases, but the liquid put into the wound when they sting often causes inflammation and painful swelling.

Fleas (Fig. 78 *A*) pass through larval and pupal stages, the larvae living in cracks in the floor. Different animals have different kinds of fleas. Monkeys, however, have none; what they are always searching for is flakes of dead skin. Rat fleas carry the germs of plague, and were the cause of the Black Death of the fourteenth century. The flea may suck the blood of a rat which has plague, and then attack a man, so passing the germs on to him. The number of fleas can be reduced by keeping houses clean, but it is really of more importance to keep down the number of rats.

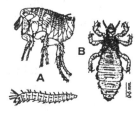

Fig. 78. *A*, Flea. The upper figure shows the Flea, the lower figure its larva. *B*, Louse.

The *Louse* (Fig. 78 *B*) spreads typhus fever. As it sucks blood, at the same time it deposits waste matter (faeces) containing fever germs on the skin of its victim. The person bitten scratches the irritating spot, and rubs in the germs. The louse attaches its eggs to human hairs. They are commonly called "Nits".

Insect pests

The number of insect pests attacking and eating farm and garden crops is very large. Black Aphids or "Blight" on beans are well known. The larvae of some butterflies, moths, flies and beetles do much damage to plants, especially those that we cultivate to eat. They may eat roots, as for instance the Carrot fly and Cabbage-root fly larvae, or they may eat the leaves, as the Colorado beetle which eats potato leaves. The plants then grow very little or may even die.

Beneficial insects

Sometimes the damage done by insects to plants is so serious that steps have to be taken to kill the pests. This is usually done by spraying the plants or the soil with chemicals. All insects, however, are not pests. Some insects benefit us by eating or killing certain pests, for instance Lady-birds eat "Blight". The most useful insect is the Ichneumon fly, which lays its eggs in or on the larvae of insect pests. The Ichneumon fly larvae live as parasites inside the bodies of their prey, on which they feed, thus gradually killing them (Fig. 79).

Parasites help to keep the numbers of the harmful pests within reasonable limits. Now, insect pests are often accidentally taken on ships from one country to another. In the new country none of their enemies may be present, so they rapidly increase in numbers. For instance, the larva of a certain moth was ruining the coconut industry of Fiji in the South Seas by eating the leaves of the palm trees. But lately a certain fly, whose larvae live as parasites in the moth larvae and destroy them, has been introduced into the island. In this way the numbers of the moth have been greatly reduced in a short time.

Fig. 79. Caterpillar showing larvae of a parasitic insect just emerging from it.

Rabbits introduced into Australia have become a plague, for there are no stoats, weasels or ferrets to kill them off. This is another example of what happens when the balance of nature is upset.

Social insects

Bees, wasps and ants are called social insects because they live together in communities. The young ones do not leave their parents, but work together to bring up further generations.

The Bee

Bees usually make their nests in hollow trees, but **hives** made by man are more convenient places for them. Nearly all the bees in a hive are **workers** (Fig. 80 *a*). These are really females that will never lay eggs. Only one female in the hive can lay eggs; she is called the **queen**. There are also male bees, called **drones**. The workers do all the work except that of laying the eggs.

When bees hatch out of the egg they start life as white legless larvae (Fig. 80 d), which turn into pupae. The pupae have the beginnings of legs, wings and eyes. Three weeks elapse before they come out of the pupal stage. The pupae turn into queens, workers or drones. A queen may live five years, but in summer workers die after eight weeks.

Inside the hive is a **comb**, which consists of six-sided cells made by the workers of wax, with passages between them. The queen lays an egg in some of these cells. Other cells are stores for food, either honey or pollen.

A worker bee does all the various jobs in the hive one after the other as she grows older. First she cleans out the cells from which new insects have emerged, so that the queen can lay another egg there. Then she feeds the helpless larvae with honey and pollen from the stores. After this she sucks nectar from the mouths of older workers; inside her body the nectar is changed into honey, which she squirts into the store cells. She also packs pollen into pollen cells, and carries any dirt out of the hive. Then she begins to build new cells with wax which is formed (secreted) in her abdomen. During this period of her life she flies also out of the hive occasionally, each time going farther afield, until she knows the landmarks in the neighbourhood. When five weeks old she is a doorkeeper, smelling bees with her feelers and driving off strangers. Finally she goes out and gathers nectar and pollen from the flowers. For carriage the pollen is packed into sticky balls on the hairs of her hind limbs.

When the hive is getting overcrowded, a few larvae are specially raised by the workers to become new queens. Extra large cells are made, in which eggs destined to become females are laid. The larvae are given food with more pollen than usual in it. This turns female larvae into queens, instead of into workers. Just before the new queens emerge from their pupae, the old queen flies out of the hive with half its workers. This is called a **swarm**. If a bee keeper is quick enough he gets the swarm to enter an empty hive.

Then a new queen emerges in the old hive, and the colony starts off afresh with its new queen. She begins her career by flying out on a wedding flight followed by the drones. On the

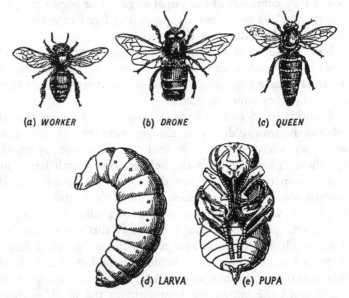

(a) WORKER   (b) DRONE   (c) QUEEN

(d) LARVA   (e) PUPA

Fig. 80. Different types of hive bees. Larva and pupa of Bee.

return of the newly wedded queen, the workers kill the other queen pupae. In the autumn the drones are all turned out of the hive and either die of starvation or are stung to death by the workers.

Ants

As in the case of bees, there are male, female, and worker Ants (Fig. 81). The social life of ants, however, is more complicated even than that of bees. The worker ants do one job only. Some kinds of ants bring pieces of leaves into their nests and manure them with their own waste matter or *faeces*, so that fungi grow on the leaves. The ants enjoy eating these fungi.

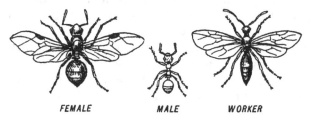

FEMALE   MALE   WORKER
Fig. 81. Wood Ants.

They also keep "cows". Certain plant lice (aphids) produce a sugary liquid which ants like, and the ants tickle them to get it. Other ants capture the workers of different kinds of ants and make their captives work for them as slaves.

Chapter 9

ANIMALS WITH BACKBONES

All animals with backbones belong to one of five classes: 1, Fishes; 2, Amphibians; 3, Reptiles; 4, Birds; 5, Mammals.

Fishes

These animals are clearly adapted for life in water. If we look at a Goldfish swimming, for example, we see that the cylindrical body which is covered with scales, and the pointed head and tail, allow the fish to swim without any difficulty. In fact the body is *stream-lined*. Inside the animal there is an *air bladder* which makes the fish float more easily and so it needs to make no effort to keep up in the water.

Fishes have *fins* for swimming (Fig. 82). If you look at a Goldfish you will see that it has two pairs of fins which correspond to our arms and legs, and three simple fins, the tail fin, the dorsal fin on its back and the anal fin near to the anus.

Fishes have *gills* for breathing instead of lungs like we have. Blood containing carbon dioxide is sent by the heart to the gills. Here carbon dioxide passes out in solution into the water, and dissolved oxygen is taken in from the water into the blood, in which it circulates to all parts of the body. The gills are protected on each side of the body by a *gill cover*. If you lift up the gill

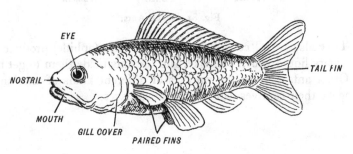

Fig. 82. The Goldfish.

cover of a Herring you will be able to see the structure of the gills. The spaces between the gills, called *gill slits*, open from the mouth to the space under the gill cover. When a fish opens its mouth to breathe, water enters the mouth and at the same time the gill cover is closed. Then the mouth is shut, driving the water through the gill slits and past the gills, while the gill cover opens as the water is forced out.

I expect you have all heard of, or eaten, "hard" and "soft" roe of fishes. The "hard" roe is a mass of small eggs which the female lays in the water. The "soft" roe is a mass of sperms (see Chapter 7) which the male sheds on top of these eggs, when they have been laid, to fertilise them. Most fishes do not protect their eggs after they have been laid. Instead, they lay a large number of eggs. Then in spite of the dangers that surround the young fishes, it is probable that at least a few of the many young will survive until they are fully grown.

Some fishes, however, do protect their eggs. The small fishes with spikes on their backs found in ponds and streams which are called Sticklebacks build nests of weeds. The male builds the nest. When eggs have been laid in the nest by several females and fertilised by the male, he keeps guard over the nest, fiercely driving away intruders.

The Salmon

Both the Salmon and Eel are interesting because they pass part of their lives in fresh and part in salt water.

The largest Salmon are from 4 to 5 feet long. They are found on both sides of the North Atlantic and the North Pacific Oceans. When large Salmon, which are living in the sea, are ready to lay their eggs, or *spawn*, they travel inland up the rivers, often jumping up waterfalls or weirs. The eggs are laid in rivers during the months of September to January. When the young fish are about two years old, they travel down to the sea, where they feed and grow rapidly. One or two years afterwards they return to the river to spawn.

The Eel

Eels living in the streams and rivers of all countries of Europe go down to the sea to breed. They travel to the depths of the Atlantic Ocean near the West Indies. Here they spawn, but never return again. The young Eels come back to the European rivers, taking about three years to do so, having travelled at the surface of the sea.

Flat fishes

Plaice, Soles and Turbot are peculiar fishes living in the sea. If you look closely at them you will see that they are lying, and they swim, on one side of their bodies. The gill openings are one on the upper, and one on the lower surface. The mouth is placed as it would be if the fish swam like other fishes, but the skull is twisted, so that both eyes are on the upper side (Fig. 83). These Fishes start life swimming as other Fishes do, but later they turn on one side, while the skull gradually twists. They can change their colour and pattern to match the ground on which they are lying.

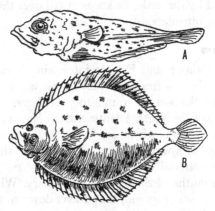

Fig. 83. Flat fish. *A*, Young stage which looks and swims like an ordinary fish. *B*, Adult stage [drawn to smaller scale]. Both drawings are of the left side of the fish.

Amphibians

Frogs, Toads and Newts belong to this group. They all start life as water animals and then change into land animals. Most of you, I expect, have watched Frogs gradually develop out of the eggs. During the early spring you will find **Frog spawn** in most ponds, ditches and canals. The spawn consists of a number of eggs. Each egg is surrounded by clear jelly which swells when laid by the frog in the water (Fig. 84 *A* and *B*). If you get some Frog spawn and put it into an aquarium or a glass jar containing water and water weeds, you can watch the eggs gradually growing and changing into Frogs (Fig. 84).

Life history of the Frog

Several days after they have been laid, the black eggs change their shape (Fig. 84 C and D), until after about 14 days a tiny creature emerges out of the jelly (Fig. 84 E). This small animal has no mouth, no eyes, no gills and no limbs, but only a **sucker** with which it fastens itself to a leaf (Fig. 84 F). It is fed by the

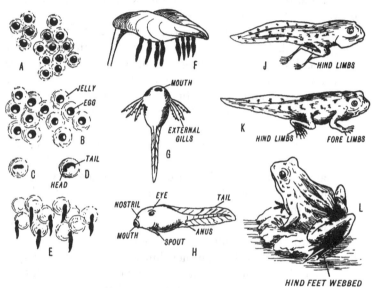

Fig. 84. Life history of a Frog.

food, or yolk, which still remains inside it from the egg. A **mouth** soon develops in front of the sucker. The small animals, or **Tadpoles** as they are called, breathe by means of gills, called **external gills** because they grow outside the body (Fig. 84 G). Through the gills oxygen dissolved in the water is taken into the blood and carbon dioxide is passed out in solution into the water. As the Tadpole increases in size four **gill slits** appear on each side of the neck, and **internal gills** are formed like those in Fishes, while the external gills shrivel. A fold of skin forms a gill cover over the internal gills, and the Tadpole then breathes like a Fish (Fig. 84 *H*). Water goes into the mouth, passes over the gills, and is then forced out through one opening called the **spout** (Fig. 84 *H*). When the Tadpole is fully grown the **back legs** appear, and several days later the **front legs** appear, one emerging through the spout. **Lungs** develop while the gills wither and the tail shortens and then the young Frog comes out of water on to the land (Fig. 84 *L*).

At first Tadpoles live on plants, but as they turn into Frogs they eat small Insects. Fully grown Frogs eat Insects, Worms or Slugs. The tongue of a Frog is peculiar, as it is joined to the front and not the back of the mouth. The Frog can throw its tongue forward, and then the tip reaches a long way beyond the mouth. This is useful when the Frog is catching insects with its tongue.

Tadpoles are readily eaten by many animals in ponds, and Birds and Snakes eat small Frogs. So Frogs, like most Fishes, lay many eggs, so that at least a few of the young survive.

Frogs have a naked skin, that is the skin is not covered with scales, feathers, or hair. Our skin is covered by an outer layer of dead skin. This is not so in Frogs. The outside of their skin is made up of living cells, so they are able to breathe through their skin as well as through their lungs. If their skin becomes dry, the cells die and they can no longer breathe through their skin, so they die. Frogs can be compared in this respect with Earthworms (Chapter 7).

During the winter, when the temperature is low, Frogs **hibernate**, that is they hide themselves in holes in the ground where it is damp. Here they remain, almost lifeless, until the springtime.

Toads

Toads can be distinguished from Frogs by their shape and their skin which is more wrinkled and warty and less moist. Their eggs are laid in long strings of jelly, which are fastened round water weeds. Each string of jelly contains 20 to 30 eggs.

Newts

Compare the life history of a Newt (Fig. 85) with that of a Frog. The eggs are very difficult to find. They are laid singly on leaves of water plants, the edges of which are carefully folded

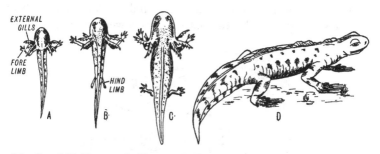

Fig. 85. Life history of a Newt. *A, B* and *C* are stages in development. N.B. the fore limbs appear before the hind limbs. *D,* Adult.

over. If you are fishing in a pond or ditch you may see young Newts amongst the weeds. At first you may think that they are small Fishes, but on looking closer you will see the external gills and tiny limbs (Fig. 85 *A, B* and *C*). The limbs develop very early in life, and the young Newt keeps its external gills for a long time. Unlike the Frog, the fully grown Newt does not lose its tail. During the breeding season the male Newt has a crest along the middle of its back.

Reptiles

Snakes, Lizards, Crocodiles and Tortoises belong to this group. The skin is covered by horny scales in Lizards and Snakes, and by an armour of thick scales in Crocodiles and Tortoises. Most Reptiles hatch out of hard-shelled eggs, the young ones being similar in appearance to their parents. The eggs of the Adder hatch before they are laid, so the young ones are born alive. There are between five and six thousand kinds of Reptiles in the world to-day. As they prefer a sunny climate, only six kinds are found in Britain, three sorts of Snakes and three kinds of Lizards.

Snakes

Many Snakes have a bite which is poisonous. Small channels connected with poison glands are found in the front teeth or "fangs" (Fig. 86). The two halves of the lower jaw are not firmly

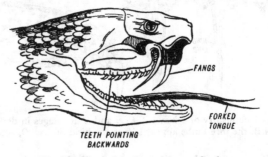

Fig. 86. The head of a poisonous Snake.

fixed together in front as ours are, but are joined by a substance which can stretch like elastic. This allows the Snake to swallow prey larger than it could otherwise do.

The **Adder** or **Viper** is the only poisonous Snake found in Britain and it feeds on Frogs and other small animals. The harmless **Grass Snake** feeds on Frogs and Fishes, and can sometimes be seen swimming about in water. The colour of the Grass Snake varies. Some are almost black, but usually they are a grey-green colour. The underparts are black, and behind the head, which is also black, there is a yellow collar. The Adder does not have a yellow collar behind its head. It is usually a dark-reddish brown colour, and along its back is a line of black blotches joined together to form a zig-zag pattern. Near the head this line ends, and on the head itself there is a V-shaped mark.

The three British **Lizards** are all about six inches long but are different in colour. They feed on Snails, Worms and Insects.

Crocodiles and **Alligators** have long bodies and tails covered with thick scales. Their legs are short, the tails being used for swimming. They can lie in the water, showing only their eyes and nostrils. These animals will eat Fish, but if they are hungry they seize any animal that comes near them, whether it is large or small. The female lays her eggs on the bank, then she covers them up with soil and leaves them to hatch in the sun.

Fig. 87. Tortoise.

Tortoises like most other Reptiles are found in the hotter countries of the world. The body of a Tortoise is covered with a peculiar case (Fig. 87). **Turtles** differ from tortoises in having paddle-like limbs. They live chiefly in the water, but in the breeding season the female comes to land to lay her eggs. Most of these animals are vegetarians. The best tortoiseshell is obtained from a certain kind of Turtle.

To-day Mammals are the most abundant large animals in the world. Many ages ago, long before there were any Men on the earth, this was not so. Most of the larger animals were then Reptiles, but they were different from the Reptiles alive to-day. Some of them were very large animals. In the Natural History Museum, London, the skeletons of many of these Reptiles can be seen. We know that these Reptiles existed because remains of their bones, teeth and scales have been found in the rocks. Such remains are called **fossils**.

Birds

Birds are found in all parts of the world. The structure of Birds is specialised in many ways to enable them to fly. I expect all of you when you have had a Turkey, Fowl or some other Bird for dinner have noticed that the breast bone has a ridge or **keel** down the middle of it. To this bone the wing muscles, used in flight, are attached. Birds like the Ostrich, which are running Birds and do not fly, have no keel bone. You have, no doubt, also noticed how light Birds' bones are. Many of them are hollow, and the larger ones are porous. Birds can actually pass air into their bones from their lungs, so making their skeleton lighter.

Much strenuous work has to be done by the muscles of Birds when in flight, so a lot of energy is needed. Heat and other sorts of energy are produced in animals by the chemical combination of their food with oxygen which they breathe in. As Birds require so much energy when flying, their lungs are comparatively larger than ours. In addition a number of thin-walled air-sacs, which are prolonged into several of the bones, are connected with the lungs, and serve as storehouses for air which is passed into them from the lungs. The Birds can use this air when it is needed.

Birds are covered with **feathers**, which like the hairs of Mammals are produced by cells in the skin. Between the feathers (or hairs in Mammals) and the skin there is a layer of air. This layer of air kept by the feathers prevents loss of heat from the body. Each of the large or **quill** feathers has a central quill. On either side of the quill there are **barbs** each of which has numerous branches called **barbules** (Fig. 88). These barbules have hooks on them, which interlock with the barbules of the next barb. This gives greater resistance to the air as the Bird is flying, in just the same way that we get resistance to the water when swimming by keeping our fingers together (Fig. 88).

Fig. 89 shows the structure of the wing of a Bird, and also of a Bat (a Mammal) which, you will see, is constructed in quite a different way. We all know what a variety of colours there is in the feathers of Birds, especially in those of Birds living in

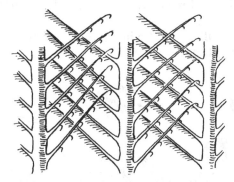

Fig. 88. Part of feather magnified, showing barbs and barbules.

Fig. 89. Wings. *A*, The Bat's wing is a skin supported by arm, second to fifth fingers, and leg. *B*, The feathers of the Bird's wing are on the arm and enlarged second finger.

hotter climates. The feathers or **plumage** of the male Birds are often more brilliantly coloured than those of the female, for example the Peacock and Peahen.

The skull of a Bird ends in a hard *beak*. The beaks of Birds vary in shape, as they are adapted to catch or eat the particular food that the Bird requires. In Fig. 90 a few different types of beaks are shown.

The feet of Birds are also interesting, as they, too, tell us the habits of the Birds (Fig. 91).

Fig. 90. Types of beaks. *A*, Sparrow: short, strong beak for eating seeds. *B*, Blackbird: slightly longer beak for catching and eating Worms and Insects. *C*, Duck: broad, flat beak for straining small creatures from the mud. *D*, Heron: long beak for catching Fish. *E*, Parrot: strong, hooked beak for holding and cracking nuts. *F*, Eagle: strong, sharp, hooked beak for catching and tearing prey to pieces.

All Birds lay eggs. These vary in size, shape and colour. Birds which nest in holes lay white eggs, whilst eggs that are laid in the open are coloured and so are not easily seen.

The majority of Birds build nests in which to lay and hatch out their eggs, though a few like the Cuckoo use the nests of other Birds. Nests vary considerably in their shape, construction and material. There is not sufficient space here to describe the many different kinds of nests, but if you look carefully you will be able to find out for yourself what materials are used. For instance a Rook makes a very coarse nest of large twigs; a Robin makes a very soft nest chiefly of moss and feathers; a Thrush makes its nest of small twigs and grass, and lines it with mud which it carefully smooths out with its breast. It is also interesting to find out in how many different places nests are found. Just a few examples can be given here. Nests are found in trees (Rooks), bushes (Thrushes), on the ground (Skylark),

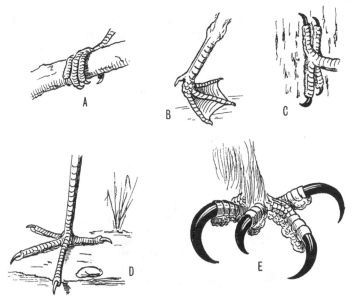

Fig. 91. Types of feet. *A*, Dove: adapted for perching with three toes pointing forward and one backward. *B*, Duck: webbed foot for swimming. *C*, Woodpecker: climbing Bird with two toes pointing forward and two backward. *D*, Heron: the spreading toes which prevent the wading bird from sinking in the mud. *E*, Eagle: the sharp curved talons of the Bird of prey.

in hollow trees (Woodpecker), under the eaves of houses (Sparrows), in holes in the ground (Sand Martin), and on the cliffs (Seagulls).

The mother bird, sometimes assisted by the father, sits on the eggs for a time until they hatch out. Some young Birds are able to run about and feed themselves shortly after they come out of the eggs. Chickens, Partridges, Grouse and Pheasants are able to do this. Young Thrushes and Sparrows, however, are very feeble when they hatch out. These young Birds are fed by their parents until they are able to seek their own food.

Migration

During the late summer or early autumn many Birds fly to other countries. We say that these Birds *migrate*. The Swallow, Swift, Cuckoo and many other Birds fly from our country to

Fig. 92. Penguins.

hotter climates. Most of them go to various parts of Africa. Other Birds, such as the Snow-bunting and Redwing, come to Britain in the autumn from the colder countries near the Arctic Circle.

Birds can be put into two classes. Most Birds are called *flying Birds* because they can fly. The Penguin (Fig. 92) belongs to this class although it can no longer fly but uses its wings for swimming. A few Birds, such as the Ostrich, cannot fly and so are called *flightless Birds*. These Birds can run away very quickly if threatened by any danger. Ostriches are so strong that they are often used to draw carriages in which one person can sit.

Mammals

This group of animals includes the largest animals living to-day. Man is a Mammal and so are Dogs, Rabbits, Horses, Elephants and even Whales. The name Mammal is given to the class because the females produce milk to feed their young. All Mammals breathe air, by means of lungs, throughout their life. The most obvious difference between Mammals and the other groups of backboned animals is that they have hair on their bodies. Mammals and Birds are called "warm-blooded" animals. Reptiles, Amphibians, Fishes and all backboneless animals are called "cold-blooded". These terms are not really true. "Warm-blooded" animals are really able to regulate the temperature of the body, so that it remains constant whether the temperature of the outer world is hot or cold. "Cold-blooded" animals have no such control, and their temperature varies with that of the outside world.

In Chapter 10 we shall carefully study the structure of the Human Body. As Man is a Mammal, we need not deal here with the structure of a Mammal's body.

There are a number of groups of Mammals.

1. Pouched Mammals

The most familiar example of this group is the **Kangaroo**. The young ones are sheltered, for

Fig. 93. Rock Wallaby.

some time after they are born, in a pouch formed by a fold of skin on the underside of the mother's body (Fig. 93).

2. Whales

Although **Whales** live in the sea and look like Fishes, they are Mammals (Fig. 94). Their bodies are not covered with scales, and they have lungs and breathe air. Their tails are flattened horizontally instead of vertically like those of Fishes, and help in the continual journeying to the surface of the water to breathe. Although these animals live in cold water, their bodies are kept warm by the layer of fat, called **blubber**, which is under the skin. Whales up to ninety feet long have been caught. They are the largest animals living. Elephants are the largest animals on land, and Giraffes are the tallest.

Fig. 94. Killer Whale.

Whales can be divided into two groups. **Toothed Whales** including the **Porpoises** and **Dolphins** which feed on Fishes, and **Whalebone Whales** which have no teeth even when they are fully grown. Instead they have hundreds of strips of whalebone forming a sieve in their mouth. Water, which is taken in with each mouthful of food, can be drained off and passed out of the sides of the mouth, whilst the minute creatures, which the Whale eats, are kept in the mouth, and then swallowed.

3. Hoofed animals

These animals are so called because each toe is protected by a horny covering known as a **hoof**. The number of toes varies in the different animals. The **Horse**, for example, has only one toe and so is said to be "odd-toed". Other animals have two or four toes and so are "even-toed". There are generally four toes, but two may not reach the ground, and appear as two

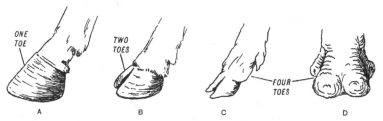

Fig. 95. Hoofed animals. *A*, Horse. *B*, Cow. *C*, Pig. *D*, Hippopotamus.

horny lumps. The **Cow**, **Goat** and **Sheep** have two toes, the **Pig** and **Hippopotamus** have four toes (Fig. 95).

Pigs will eat flesh as well as vegetable foods. All other animals belonging to this group are **herbivorous**, that is they eat only plant food. **Deer**, **Oxen**, **Giraffes**, **Camels**, Sheep and Goats

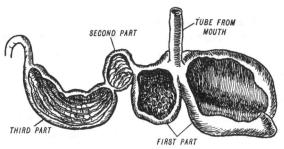

Fig. 96. Cow's stomach.

chew the cud. These animals have a very peculiar stomach (Fig. 96). Grass or other plants are quickly cropped by these animals, and swallowed. The food goes into the first part of the stomach, and remains there until the animal finds a nice quiet place where it can lie down and chew its meal at leisure. Sodden food is then passed back in small quantities from the first part of the stomach into the mouth again, where it is well chewed, ground by the teeth, and mixed with a juice called **saliva**, produced in the mouth. The food is then swallowed a second time and passed into the second part of the stomach, where it is strained. It then goes into the third part of the stomach where digestion continues.

4. Elephants

Elephants are the largest land animals living to-day. Although an Elephant is so large it can walk along making very little noise, as it has padded soles on its feet. Elephants are peculiar in having **tusks**, which are very long teeth, and also a **trunk** which is a drawn-out snout. The Elephant uses its trunk for a variety of purposes—for putting food into its mouth, rooting up plants, feeling its way in the dark, and sucking up water for spraying over its body or for drinking. Elephants are herbivorous.

There are two kinds of Elephants—the African Elephant and the Indian Elephant. The Indian Elephant can be distinguished from its African cousin by its larger head and smaller ears and tusks. This Elephant can easily be taught to do useful work for Man. The African Elephant is very difficult to tame, and is valued more for its ivory than for its usefulness.

5. Beasts of prey

All those Mammals which prey on or kill other animals for their food are included here. These flesh-eating animals are said to be **carnivorous**. They all have sharp claws on their limbs. Animals similar to the **Cats**, **Lions** and **Tigers** are able to hide their claws inside their feet when walking or running (Fig. 97). This enables them to move about silently, and the claws are also kept sharp. **Dogs**, **Wolves**, **Foxes** and **Jackals** cannot withdraw their claws into their feet, so the claws become blunt and cannot be used for catching prey. These animals run down their prey using only their teeth for killing.

Carnivorous animals always have large, strong and pointed

side or *canine* teeth which are used for catching prey. They also have sharp pointed back teeth which are used for tearing food to pieces.

A Cat has a rough tongue with which it can remove flesh from bones. A Dog has a smooth tongue, and so has to bite the meat with its teeth.

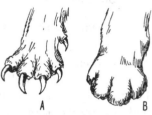

Fig. 97. Cat's feet. *A*, showing claws. *B*, claws hidden.

Weasels are very useful beasts of prey as they prevent the country-side from being overrun by Rabbits, which they kill.

Bears also belong to this group. The white Polar Bears live chiefly on Fish. The Brown Bears are not entirely carnivorous

Fig. 98. The Sea-lion. Limbs adapted for swimming.

as they will eat fruit, berries, and other plant food and are very fond of honey.

Seals and *Walruses* live in the sea, their food consisting of Fish. Their front and hind limbs are peculiarly shaped, and can be used for swimming, as well as for "waddling" along the ground (Fig. 98). The baby Seals are helpless when they are born, but they soon learn to swim.

6. Gnawing Mammals

This is a large group of fairly small Mammals, which are chiefly vegetarians. They have many enemies, and cannot protect themselves very well, but they are found in large numbers because they breed very quickly. **Rabbits**, **Hares**, **Rats**, **Mice** and **Squirrels** belong to this group. They have front teeth which grow continuously throughout their life, as they are always being worn away by constant nibbling of food.

Rabbits make burrows in the ground, but the Hare never burrows. Young Rabbits are very helpless when born, and have only a very sparse covering of hairs. The mother puts them into a nest which she makes with hair off her own body.

A Squirrel can climb trees and has a bushy tail. It hibernates during the winter.

Mice and Rats are pests to Man. Certain rats, which are sometimes brought to our docks, carry the germs of plague. Circular discs are put on the ropes of ships in dock, to prevent Rats entering by running up the ropes.

7. Insect-eaters

Hedgehogs, **Moles** and **Shrews** belong to this group of Mammals, which is another large group of small animals. Although they are called Insect-eaters, these Mammals will eat other things as well, such as Mice, Snails, Worms and Slugs. Hedgehogs are also fond of eggs. Moles eat Earthworms which they find in the soil where they live. The Hedgehog hibernates through the winter, living on the fat which is stored in its body.

8. Bats

Bats are very similar to the Insect-eating Mammals, but they are put into a group by themselves because they can fly. Each wing is made of skin which stretches between the four very long fingers of the hand, and the side of the body. The thumb alone is not part of the wing, and ends in a hook-like nail. When asleep, Bats hang downwards by the claws of their feet (Fig. 99). They sleep in a dark corner during the day. Bats have very poor sight, but the skin on the face is very sensitive to the movement of the air, so that they can avoid obstacles even when flying in darkness.

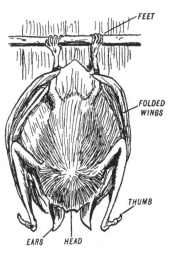

Fig. 99. A Bat asleep.

9. Men, Apes and Monkeys are called Primates

All these animals have a fairly large brain, so their heads are different in shape to other Mammals. Their eyes are at the front of the head, and not at the side. Except for **Man**, these animals are adapted for climbing trees, and their feet, as well as their hands, can be used for grasping. The Man-like Apes, such as the **Gorilla**, **Orang-utan** and **Chimpanzee** can walk on their hind limbs as we do. The Monkeys of America possess tails with which they can grasp, and these are useful when they are climbing. These animals are mostly vegetarian.

Chapter 10

HUMAN PHYSIOLOGY

The structure of the bodies of Apes and Monkeys is very similar to that of our own. In fact the bodies of all Mammals are built on the same plan, but those of Monkeys are most like ours.

The skeleton

The skeleton is the bony framework of the body. It consists of about two hundred bones which vary in size and shape (Fig. 100). These bones can be divided into four groups: (1) The skull. (2) The backbone or vertebral column. (3) The shoulder girdle and arms. (4) The hip girdle and legs (Expt. 2, Chap. 10).

The *skull* consists of the brain case and the face. The brain case is a hollow, bony case inside which is the brain. The bones forming this case are joined together by irregular saw-like edges (Expt. 1, Chap. 10). In young children the bones are not firmly joined together, so they have room to grow. All the bones of the face, except the lower jaw, are joined to the brain case to form one solid piece. The lower jaw is jointed to the skull just in front of the ears. It is fixed in such a way that it can move downwards, upwards, sideways, backwards and forwards. The teeth are fixed in the upper and lower jaws. At the base of the skull there is a large hole, where the spinal cord (see p. 123) passes out of the brain. On either side of the hole there is a large round knob which fits into a hollow in the first bone of the spinal column.

The *spinal column* is the axis with which all parts of the skeleton are connected. It is made up of thirty-three small bones called *vertebrae*. The first twenty-four are jointed in such a manner as to be movable on each other. The last nine are joined

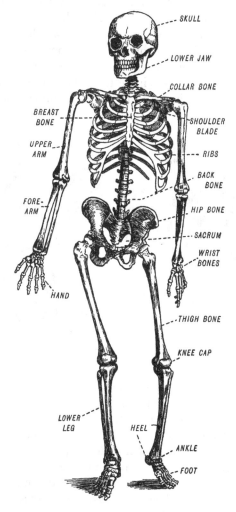

Fig. 100. The human skeleton.

together. There are seven vertebrae in the neck. (There are only seven even in the Giraffe's long neck.) The next twelve vertebrae belong to the back and support the **ribs**. The remaining five movable ones belong to the loins. The next five vertebrae are fused together to form a solid mass of bone called the **sacrum**. The last four bones are imperfectly formed and are joined to the sacrum. These bones form the tail in other Mammals.

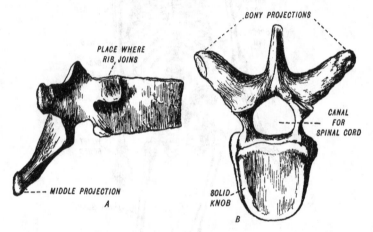

Fig. 101. A vertebra. *A*, Side view, right side. *B*, View from above.

A vertebra consists of a round, solid knob with a bony ring towards our back enclosing a space (Fig. 101). The spaces of all the vertebrae are connected with one another, forming a channel through which the spinal cord runs. Three pieces of bone grow out from the bony ring. The middle one projects backwards, and can be easily felt through the flesh on the back. The ligaments which hold the bones firmly together, and the muscles which allow us to bend our backs, are fastened to these bony projections. Between the bodies of the vertebrae are pads of an elastic, gristle-like substance called **cartilage**. These pads unite the vertebrae, and also make the backbone flexible.

The **shoulder girdle** consists of two **collar bones** at the front, and two **shoulder blades** at the back. The collar bones, which are narrow bones, can be felt at the base of the neck. Each one is joined to the top of the **breastbone** at one end, and to the shoulder blade at the other. The collar bones are very short and imperfect in the Cat and Dog, and are missing altogether in the Horse and Sheep. They are, however, well developed in climbing animals.

The shoulder blade is triangular in shape (Expt. 3, Chap. 10). It consists of a broad, flat part, with a prominent ridge near the shoulder, which can easily be felt through the muscles. The shoulder blade is broadest towards the backbone and narrow towards the tip where it joins the collar bone. At this end there is a hollow cavity which receives the rounded head of the upper arm.

The upper **arm** is made of one bone, and the lower or forearm of two bones. There are eight bones in the wrist and five in the palm of the hand. Each finger has three bones and the thumb two bones.

We know that the twelve vertebrae of the back have each a pair of ribs jointed to it. The ribs nearer the neck are shorter than those farther away. The first seven pairs bend right down and round and join the long, flat bone called the **breastbone** which stretches downwards from the neck. The next three pairs of ribs do not join the breastbone, but finish by joining the pair above. The last two pairs are very short and do not join anything. They are called *floating ribs*. The ribs are joined to the breastbone with cartilage. The ribs together with the backbone and the breastbone form a framework with movable sides, which encloses and protects the heart and lungs.

The **hip girdle** consists of the sacrum and two large **hip bones** which together form the **pelvis**. The pelvis has to support the weight of the body. On either side of the hip girdle there is a deep hollow which receives the top of the thigh bone.

The **legs** are very similar to the arms, there is one bone in the thigh, two in the lower leg, seven bones in the ankle, five in the instep, two bones in the big toe and three in each of the others. One ankle bone is larger than the others and forms the heel. In the knee there is an extra bone, the knee cap. Without it we could not stand upright as it prevents the leg from bending backwards.

Functions of the skeleton

We have learned that the skeleton has several functions: (1) It makes the body to a certain extent rigid. (2) It protects several important organs of the body such as the brain, spinal cord, heart and lungs.

When we are very young we have scarcely any bone in our body, its place is taken by cartilage. This is why the skeleton of a child is softer than that of an adult, and is easily bent into a wrong shape. As we get older the cartilage is replaced by bone. Bone is made by living cells.

Muscles

Red meat consists of muscle, which is made up of living cells. The function of the muscle is to move the various parts of the body. The action of a muscle is easily understood if you feel the muscle in your upper arm, called the **biceps muscle**. Stretch out your right arm and hold the upper part with your left hand. Gradually bend your right arm. The upper arm gets thicker. The muscle has two narrow ends and a thicker part in the middle. The ends or **tendons**, or sinews as they are often called, are attached to bones. The upper end of the muscle mentioned is joined to the shoulder girdle, the lower end to one of the bones of the forearm. As the muscle contracts it becomes shorter and thicker and draws up the forearm (Fig. 102).

All muscles can shorten in length, or contract. The muscles of the body are divided into two groups: (1) Those attached to bones. (2) Those not attached to bones, but to the ends of other muscles. Many of the second group of muscles enclose a space.

For instance the muscles in the walls of the intestine, the heart and the bladder. The muscles in the walls of the heart contract to force the blood out of the heart (see p. 118) and then relax, so allowing blood to enter the heart.

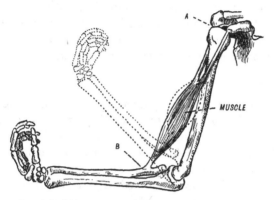

Fig. 102. Action of the biceps muscle. The muscle is attached by two tendons to the shoulder at *A*, and by one tendon to the forearm at *B*. When the muscle contracts the forearm is drawn up.

The muscles that we can move of our own free will are called *voluntary muscles*. The muscles of the arms and legs are voluntary. *Involuntary muscles* are those which work on their own, such as the muscles in the walls of the heart and the intestines.

Food and digestion

All animals must have food or else they would die. The food gives us the energy to do work, and it also makes animals grow, and helps to replace any parts of the body that are old or worn out.

Different kinds of food

All the food substances can be put into several groups: (1) *Carbohydrates* which include starch and sugar. (2) *Fats* are contained in butter, fat meat, milk, and vegetable and animal oils, such as olive oil and cod liver oil. (3) **Proteins** which are present in all plants and animals and are eaten in eggs, milk, cheese, meat, flour, cereals, and all plant foods. (4) **Vitamins** are most important to health but are only required in small quantities. Different foods contain different vitamins. (5) Certain **salts** are also necessary. Phosphate and carbonate of lime help to make bones and teeth hard. (6) **Water** is very necessary. Most foods contain a certain amount of water.

Digestion

The food which we eat goes along a tube called the **alimentary canal** or gut (Fig. 105), which passes through the body. The water and salts can pass through the wall of this tube into the blood. The other foods have first to be changed. They are broken down into a form that can pass through the wall into the blood. This process is called **digestion**. It is divided into three stages: (1) Eating. (2) Digestion proper. (3) The absorption of food by the blood.

Eating

We bite our food with our front teeth. These teeth have sharp edges for cutting into the food (Fig. 103). They each have a single "root" or fang to hold them firmly in the jaw. There are four of these biting or **incisor** teeth in the middle of the upper jaw and four in the lower jaw (Fig. 103, 1 and 2). The next tooth on each side is similar to the incisors only it is a little larger. It is called the **canine tooth**. The canine teeth of the beasts of prey are very large, strong and pointed and are used for catching prey (Expt. 4, Chap. 10). The last five teeth on each side in the upper and lower jaws are "double" teeth. The tops or **crowns** of these teeth, which is the part above the gums, are very broad and ridged, and the teeth have two or three fangs. These teeth grind up the food into small pieces so that it can be more easily digested.

A grown-up person has thirty-two teeth. Young children

have a temporary set of twenty teeth called **milk teeth**. These are gradually pushed out by the **permanent teeth**. A child of thirteen years usually has all the permanent teeth except the last four. These teeth, which are called "**wisdom**" **teeth** are often not cut until a person is twenty-five to thirty years old.

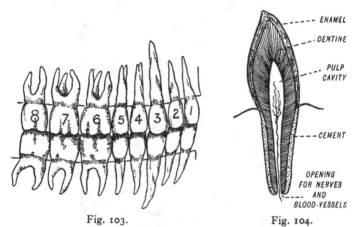

Fig. 103.   Fig. 104.

Fig. 103. The right side of the upper and lower jaw showing the teeth. 1 and 2, Incisors. 3, Canine teeth. 4, 5, 6, 7 and 8, Double teeth.
Fig. 104. Section through a tooth.

Structure of a tooth

Teeth consist chiefly of a hard substance like bone, called **dentine**. The crown of the tooth is covered by hard **enamel** which prevents Bacteria from getting to the dentine. The part of the tooth in the gum is covered by a thin layer of **cement**. In the centre of the tooth is the **pulp cavity**. Nerves and blood vessels enter this cavity through a tiny hole at the base of each fang (Fig. 104).

Digestion proper

Digestion begins in the **mouth**. There is a juice in the mouth called **saliva** which is secreted in six glands. Two of these glands are in front of the ears, and the other four are between the lower jaw and the floor of the mouth. As the food is chewed up it is mixed with the saliva, which changes the starch into sugar. Sugar is soluble (Expt. 5, Chap. 10).

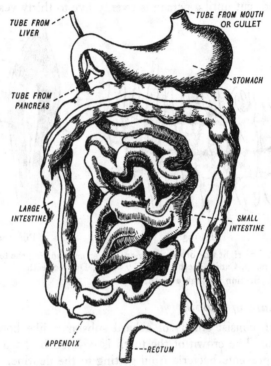

TUBE FROM LIVER

TUBE FROM MOUTH OR GULLET

STOMACH

TUBE FROM PANCREAS

LARGE INTESTINE

SMALL INTESTINE

APPENDIX

RECTUM

Fig. 105. The alimentary canal.

The food is then passed down the *gullet* (Fig. 105) into the *stomach*. A flap covers over the air tube as the food is swallowed, and prevents it from going the "wrong way". The stomach is a kind of bag with an opening at each end. Proteins are digested in the stomach by the *gastric juice* which is secreted by the inner layer of the stomach wall. The food is moved about by the movement of the stomach wall which is muscular. Some foods remain longer in the stomach than others. Pork is not readily digested, so it remains in the stomach for about four hours. The way out of the stomach is guarded by a powerful ring of muscle, which normally keeps the exit tightly closed, and opens only when the food is ready to pass out of the stomach.

The partly digested food then passes into the **small intestine**. This is a very long tube (about 20 ft. long) which is only 1 in. in diameter. It is coiled around itself (Fig. 105). In the first part of the small intestine, digestion is completed. About 4 in. from the stomach a tube enters the intestine. This tube has two branches. One comes from the **liver**. The liver secretes a greenish liquid called **bile**, which runs slowly down this tube. If there is no food to be digested the bile is stored in the **gall bladder**. Bile helps to digest fats. The other tube comes from the **pancreas**, or "sweetbread" as it is often called (Fig. 105), which also secretes a juice. This pancreatic juice can digest starches, proteins and fats.

The absorption of food by the blood

The digested food is pushed farther along the small intestine, by the contraction of the muscles of the intestine wall immediately behind the food, just as you force water along a rubber tube by pinching the tube and moving your fingers along it. The soluble, digested food then passes through the intestine wall into the blood which carries it in the **portal vein** to the liver. Here the food is further changed and then despatched in the blood to all parts of the body.

The remnants of the food, or undigested material, then pass into the **large intestine**, which is wider but shorter than the small intestine (Fig. 105). At the junction of small and large intestine a small blind pouch leads out of the large intestine and ends in the worm-like **appendix** which is about 3 in. long. The large intestine absorbs most of the water left in the waste material. The residue is then got rid of.

Respiration

In breathing the air should be taken in through the **nose**. As the air passes through the **nostrils** it is warmed in the nose. The

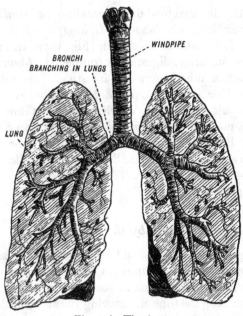

WINDPIPE

BRONCHI
BRANCHING IN LUNGS

LUNG

Fig. 106. The lungs.

larger particles of dust are taken out of the air by the hairs inside the nose. The cells in the nose secrete a sticky liquid called **mucus** which catches smaller dust particles and Bacteria. The air passes down the air tube into the chest, where the air tube forks, sending a branch to each **lung** (Fig. 106). The air tube, or "**windpipe**", is strengthened by rings of cartilage which can easily be felt in the front of the throat. When the tubes reach the lungs they divide, then each branch divides again and so on until the numerous channels are only $\frac{1}{100}$ in. across. From these fine tubes small air spaces open. This gives an enormous area of lung wall in contact with air. Very fine blood vessels, or *capillaries* as they are called, lie close to the lung wall. Oxygen passes from the air spaces into the blood, and is then carried to all parts of the body (Fig. 107). Carbon dioxide is carried from all parts of the body by the blood, and in the lungs it passes from the blood into the air spaces, and then it is breathed out.

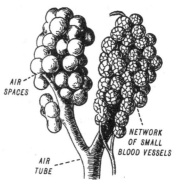

Fig. 107. Diagram showing air spaces in the lungs, and arrangement of the blood vessels round the air spaces.

The movements of respiration

The lungs are enclosed in what is almost an airtight box whose walls are movable. This box is the thorax. It is formed by the spinal column at the back, the ribs at the sides, the breastbone in front, and below by the *diaphragm*. This is a muscular partition which separates the thorax from the abdomen.

The hinder ends of the ribs are attached to the vertebral column so as to be freely movable upon it. The ribs are joined either directly or indirectly to the breastbone by cartilage, which makes the connection flexible. The ribs are normally inclined downwards and forwards. By means of several muscles the ribs can be moved from this position into one more nearly horizontal. The front ends of the ribs are moved upwards and forwards, and the breastbone is pushed slightly forward. By this movement the size of the thorax is increased from front to back. As the ribs are raised, the thorax also increases in width from side to side.

The diaphragm is normally raised up in the middle. Muscles

attached to the middle of the diaphragm stretch to the ribs and spinal column. When these muscles contract, the diaphragm flattens, so that the thorax is increased in size from above downwards.

By means of these movements of the ribs and diaphragm the thorax is increased very much in size. As it enlarges, air rushes into the lungs to distend them to the size of the thorax, that is air is inspired.

At the end of each inspiration the muscles attached to the ribs and diaphragm relax. The lungs also tend to return to their former shape, and gravity acting on the ribs tends to lower them. So air is forced out of the lungs, or expired.

The blood

Blood is a liquid that flows round the body in the **blood vessels**. The movement of the blood is caused by the **heart** which lies behind the breastbone, between the lungs (Expt. 6, Chap. 10).

The heart is a muscular organ which acts like a pump. It is made up of a right and left half which have no communication with one another. Each half is divided into two parts. The upper part of each side is called an **auricle** and the lower part a **ventricle** (Fig. 108).

Blood flows from all over the body through veins into the right auricle (blood vessels containing blood going to the heart are called **veins**, those with blood flowing from the heart are called **arteries**). As the auricle fills with blood its walls stretch. When it is full the walls contract, and the blood is forced into the right ventricle, a **valve** preventing the blood from flowing back into the vein. A valve is a covering or door which will open in one direction only. The valves or folds in the heart open when the blood flows in the right direction, but close when the blood tries to flow back again. The right ventricle contracts in turn and forces the blood into two arteries which take it to the lungs.

In the lungs the arteries divide and divide again into numerous small vessels called **capillaries**, which lie very close to the air

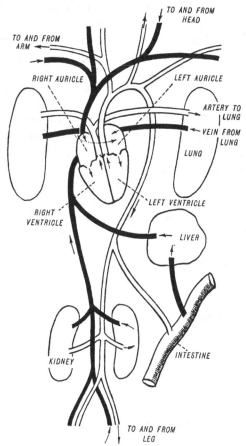

Fig. 108. Diagram of the human heart and blood vessels, seen from in front. Veins are shown black, arteries white. The valves at the openings into and out of the ventricles are indicated. The course of the circulation is shown by arrows.

spaces. Here oxygen passes from the air into the blood, and carbon dioxide passes the other way.

From each lung the blood passes along a vein to the left auricle. When full the latter contracts, driving the blood into the left ventricle. The ventricle in turn contracts forcing the blood into a large artery. This divides into smaller arteries, which again and again divide, so that the blood is driven by the heart to all parts of the body. Finally the vessels are so small that they cannot be seen with the naked eye. These are the capillaries. From the capillaries the blood gives oxygen and food to all the cells, and collects all the waste materials including carbon dioxide. Then the capillaries run together to form bigger and bigger channels, until once more they are large enough to be seen again. These are the veins which take the blood back to the right auricle of the heart.

Blood is made up of a colourless liquid called **plasma** in which a number of cells called **corpuscles** float. One kind is red in colour. These **red** corpuscles look like flattened circular discs (Fig. 109). The other kind, which are called **white** corpuscles, are very similar to the Amoeba. They are constantly changing their shape. They are not nearly so numerous as the red corpuscles (Expt. 7, Chap. 10).

Uses of the blood

The blood has a number of uses.

1. Digested food is carried to all parts of the body in the blood plasma.

2. The red corpuscles are the oxygen carriers. They can absorb forty times more oxygen than does water. Blood looks bright red in colour when it contains plenty of oxygen, and it is bluish when it contains little oxygen.

3. The white corpuscles eat up any foreign bodies, such as Bacteria, which may get into the blood. They eat up these Bacteria as the Amoeba eats its food. White corpuscles also put substances into the blood which help to get rid of the Bacteria or "germs".

4. When a blood vessel is cut, blood flows out. Death might occur if too much blood were lost. Fortunately as the blood flows out of the cut vessel it gradually becomes solid, or **clots**. Thus a plug is formed in the cut end of the vessel and so the bleeding is stopped. This clotting of the blood is brought about by a substance which the white corpuscles put into the blood.

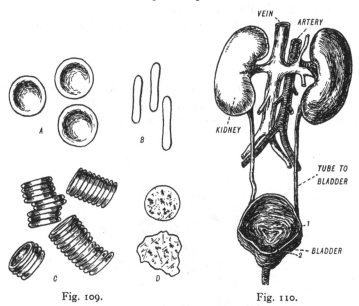

Fig. 109.   Fig. 110.

Fig. 109. Blood corpuscles. *A*, Red corpuscles seen on flat; *B*, on edge. *C*, Red corpuscles run together into rows. *D*, White corpuscles.

Fig. 110. The urinary organs. 1, Places where the tubes from the kidneys enter the bladder. 2, Tube from the bladder.

Excretion

The waste products of the body are got rid of, or **excreted**, in several ways. There are two **kidneys** placed one on each side of the "loin" region of the spine. The depressed or concave side of the kidney is turned inwards (Fig. 110). From the middle of this side a long thin tube goes down to the **bladder**. In the kidneys certain waste products are taken out of the blood. These waste products and a large amount of water pass down the tubes into the bladder, where the liquid, or **urine**, is stored. At intervals the muscles of the bladder wall contract to empty the bladder.

The bile secreted by the liver also contains some waste products. These are passed into the intestine and got rid of with other unwanted substances by the gut.

We know already that the waste product carbon dioxide is got rid of by the lungs.

The skin

The outer part of the skin consists of dead cells, below which are the living cells (Fig. 111). The dead cells are gradually rubbed off and as the living cells near the surface they die.

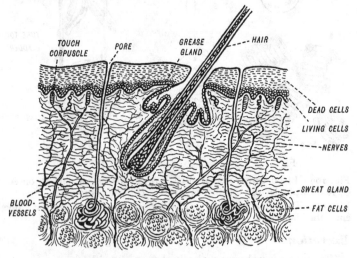

TOUCH CORPUSCLE   PORE   GREASE GLAND   HAIR

DEAD CELLS
LIVING CELLS
NERVES
SWEAT GLAND
FAT CELLS
BLOOD-VESSELS

Fig. 111. Section through human skin.

Each **hair** is formed by cells at the bottom of a deep, narrow pit in the skin. Small glands open at the base of the hair, and secrete grease. If you look at your finger tips with a lens you will see a number of openings or **pores**. These are **sweat glands**. Sweat glands secrete sweat which has been brought to it by the blood capillaries. The body loses a large amount of water daily in sweat, even when the body is at rest. In warm weather more blood goes to the skin, which perspires freely.

Temperature control

Birds and Mammals are warm-blooded animals. They are able to regulate the temperature of their body so that it remains constant. How is this temperature controlled? If the outside temperature is low less heat is given off by the skin because the blood vessels to the skin become narrow, so that less blood is taken to the skin. The fat underneath the skin, and the hair on Mammals also prevent loss of heat. Hair and feathers retain heat because they keep a layer of air next to the body which will not allow the heat to pass out. If the outside temperature rises, the blood vessels in the skin increase in width. More blood is then taken to the skin to lose heat. During hot weather the sweat glands secrete more sweat. Heat from the body is used up as the water in the sweat evaporates, so the body gets cooler. Dogs lose heat by panting, as the water evaporates from their mouth.

The nervous system

The **brain** and **spinal cord** form the central nervous system. Nerves carry impulses or "messages" either to or from the central nervous system. The nervous system forms a kind of telephone system throughout the body whereby all parts of the body are directly or indirectly in communication with each other. The **nerves**, which look like small, white threads, are the "telegraph wires" along which the messages travel. The spinal cord and the brain are the "exchanges". The nerves which carry "messages" or impulses to these "exchanges" are called **sensory** nerves. These nerves arouse in our brain the sensations of touch, sight, hearing, pain, etc. Other nerves take "messages" from the brain and spinal cord to muscles or other cells of the body. These are called *motor* nerves. Their "messages" cause the muscles to contract.

Reflex actions

If we unintentionally prick one of our fingers, or touch a very hot object, the hand is jerked away almost before we are aware of what has happened. This is a *reflex action* of the nervous system. When the stimulus is applied to the finger, a "message" is sent along the sensory nerve to the brain, and another impulse is passed out along the motor nerve to the muscles causing the hand to be jerked away.

Special sense organs

Some sensations are only felt when an impulse is applied to a particular part of the body. Thus the sensations of taste and smell are confined to certain regions of the mouth and nose; those of sight and hearing to the eyes and ears respectively, and those of touch to the skin.

Touch

There are nerves in the skin which make it sensitive. Some of the nerves end in single cells or in groups of cells called *touch corpuscles*. Pressure is felt more in some parts of the skin than in others. The skin of the forehead and back of the hand is very sensitive to pressure.

Heat

The sensation of heat or cold has to pass through the outer layer of the skin to the nerves. Different parts of the body are more sensitive to heat than others. The skin on the cheeks and palms of the hand is very sensitive.

Taste

The organ of taste is the skin or *mucous membrane* covering the *tongue* and the *palate*, which is the back part of the roof of the mouth. The tongue is joined by means of muscles to the lower jaw. The upper surface of the tongue is made rough by small lumps or **papillae** which are richly supplied with nerves. Most of the papillae are pointed. Different parts of the tongue are sensitive to different tastes. The tip is most sensitive to sweet substances, the back to bitter, and the sides to acids.

Smell

The organ of the sense of smell is the delicate mucous membrane which lines the upper part of the **nose**. You see that the organs of smell and taste are at the beginning of the alimentary canal, and give us information concerning the food we are about to eat. Hence these organs are protective.

Sight

The **eyes** are the organs of sight, which not only enable us to tell the difference between light and darkness, but form pictures of the outside world which are sent by nerves to the central nervous system.

Light passes into the eye through the hard, outer, transparent skin called the **cornea**, to the **lens**, which is a hard, oval-shaped lump of transparent jelly (Fig. 112). The lens focuses the images of all objects on the back of the eye or **retina**. On each side of the lens there is a space. The space in front of the lens is filled with a watery liquid, the space behind the lens is filled with a thin, transparent jelly. The lens is able to become fatter or thinner by means of the muscles, and this enables it to produce clear images on the retina of near and far objects.

The coloured part of the eye is the **iris** and is placed just in front of the lens. In bright light the iris covers almost the whole of the lens leaving only a small opening which is called the **pupil**. This prevents too much light entering, and so injuring, the eye. In dim light more of the lens is left exposed, and so the pupil becomes larger.

The iris of a cat's eye almost disappears at night time. It is beeause of this that a number of people have the mistaken idea that cats can see better in the dark than in daylight. All objects at which we look particularly are brought to focus on the **yellow spot** of the retina. Any object which is focused on the **blind spot** cannot be seen. Fig. 113 will show you how to prove the existence of the blind spot.

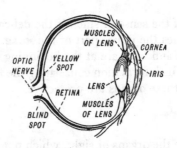

Fig. 112. A sectional diagram of the eye.

Fig. 113. Hold the book at arm's length and closing your left eye look with the right one at the left-hand spot; now bring the book slowly closer and you will find that at a certain position (roughly 8 in. away) the X will disappear from view. This is because it is focused on the blind spot.

Short sight and long sight

Sometimes the lens of the eye is not able to grow sufficiently thin to focus the image of distant objects upon the retina. Consequently the image produced is blurred. This will be readily understood by referring to Fig. 114 *a*. The defect is remedied by spectacles fitted with concave lenses.

Many people are long-sighted. This means that the lens of the eye focuses images behind the retina. Spectacles fitted with convex lenses will remedy this defect by bringing the clear image forward (see Figs. 115 *a* and 115 *b*).

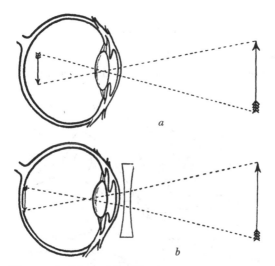

a

b

Fig. 114*a*. The cause of short sight. The image cannot be seen until it falls on the retina.

Fig. 114*b*. Short sight is remedied by viewing objects through a concave lens.

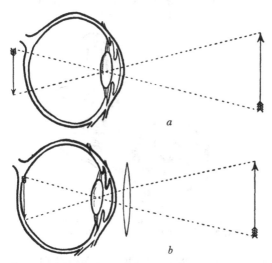

a

b

Fig. 115*a*. In long sight the lens of the eye brings images to a focus behind the retina. Consequently the retina receives a blurred image.

Fig. 115*b*. A convex lens placed in front of the eye overcomes the defect of long sight by bending the rays inwards a little before they enter the eye, thus bringing the image forward.

Hearing

The **ear** consists of three parts: the outer, middle and inner ear (Fig. 116).

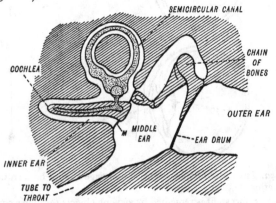

Fig. 116. The ear. *M*, The membrane between the middle ear and the inner ear.

Fig. 117. Interior of the cavity of the left inner ear. 1, 2 and 3, Semicircular canals. 4, Cochlea.

The **outer** ear consists of the part which projects from the side of the head, and the **canal** which is about 1¼ in. long. The skin at the outer end of the canal is provided with small hairs, and there are also glands, similar to the sweat glands, which produce **wax**. Both the hairs and wax prevent any dust or insects from getting into the ear.

The **middle** ear is separated from the canal by a membrane called the **ear drum**. The middle ear is almost surrounded by a bony wall. A tube, about 1½ in. long, passes from the middle ear to the upper part of the throat. Air passes along this tube into the middle ear, so that the air pressure is equalised on the two sides of the drum. If the tube becomes blocked, as it sometimes does when you have a cold, you cannot hear well.

The **inner** ear is a very complicated cavity hollowed out of bone, and separated from the middle ear by a membrane. A chain of three bones connects the ear drum to this membrane. The cavity consists of two parts (Fig. 117). The semicircular canals (Fig. 117, 1, 2 and 3) are the organs which enable us to keep our balance. The cochlea (Fig. 117, 4) is the part where the impulses are received by the nerve endings. Both parts are filled with a liquid.

The impulses giving rise to the sensations of sound are due to the movement of air particles. These impulses make the ear drum vibrate, then the chain of bones, and finally the membrane between the middle and inner ears and the liquid in the cochlea. The impulses pass from the liquid to the nerve endings and then to the brain.

Chapter 11

HYGIENE

A healthy body is much more valuable to us than anything else. Without good health we cannot do our work properly and we cannot enjoy our leisure time.

If we wish to remain healthy there are several things that we must remember.

1. The body needs fresh air and sunlight.

2. Cleanliness of the body, clothes, food and surroundings is most important.

3. The body must have exercise and rest.

4. Wholesome food is necessary to the body.

5. You must take all the care possible to prevent the body from being harmed in any way.

1. Fresh air

Life lived in the open air and sunshine makes stronger men and women than life lived indoors. We cannot all lead an outdoor life, but we can spend as much time as possible in the open. When we are compelled to stay indoors, we should see that the air in a room is kept fresh. With proper ventilation the air in a room is kept constantly moving. This is necessary for health. In still, moist, warm air a person loses vitality and may become ill.

It is important that a person should breathe deeply. Only in this way can all the air spaces of the lungs be properly filled with air.

Sunlight

Sunlight gives health and strength to the body and helps to destroy the Bacteria which cause diseases. The more sunlight that we get on our bodies the better. We must, of course, be careful that we do not burn the skin. Also the head and nape of the neck must be protected or we are likely to have sunstroke.

2. Cleanliness

"Where there's dirt there's danger" is a slogan of the Health and Cleanliness Council, an organisation formed to assist all those people teaching hygiene in schools. This slogan is very true. Uncleanliness is the cause directly or indirectly of many ailments and diseases.

Personal cleanliness

The skin. After studying the structure of the skin it is very easy for us to understand why it is so important to keep the skin clean. First of all the sweat comes out of the pores and dries on the skin. Dust and Bacteria soon collect on the dried perspiration. The perspiration, dust and germs must be removed by washing the skin with warm water. Cold water acts as a tonic to the skin. The whole body should be washed at least once or twice a week, and the exposed parts such as the face and hands must be cleaned frequently.

If dirt is allowed to collect on the skin the pores become blocked, and the body cannot perspire. Any Bacteria on the skin increase in numbers in the dirt and may give rise to a skin disease. Furthermore if the skin is injured, it heals more quickly when it is clean.

The hair

The hair should be well looked after. It does not require washing too often, but it should be brushed and combed frequently. Sores may develop on the scalp if it is dirty. The Head Louse and its eggs, called "Nits" (see Chapter 8), will not be found on hair that is well looked after.

The nose

The nose must be kept clean and clean handkerchiefs must frequently be used. The hairs inside the nose gradually become blocked by the dust particles which they remove from the air that is taken in. Unless the nose is clean, air containing dust and Bacteria will pass down to the lungs. Children often develop the habit of breathing through the mouth, as a result of not keeping the air passages clear. This is very important if a child has nasal catarrh or a "cold in the nose". Air must be taken in through the nose so that the dust and Bacteria are removed, and the air is warmed before reaching the lungs.

The ears

Ears must be kept clean. Wax sometimes collects in the ear and causes a person to be slightly deaf. This wax should be removed by a doctor, as an inexperienced person could easily damage the ear drum. No solid body should be put into the ear, as the ear drum may be damaged. A blow on the ear is very harmful, as it may cause a person to become deaf. If a discharge comes from the ears, a person should have proper treatment at once.

The eyes

The eyeball keeps itself clean. The tear gland on the outer side of the eye secretes a watery liquid which washes away any small object that may get into the eye. The liquid then passes down the tear duct into the nose. The eyelids must be kept clean or else they become red and sore.

The mouth

It is most important that the mouth should be kept clean as it is the gateway to the interior of the body. The teeth must be well looked after as they play an important part in the digestion of food.

Teeth must be cleaned regularly, as food and Bacteria collect in the spaces between them. They should be cleaned every night and morning, and if possible after every meal. If food remains in the mouth, the enamel of the teeth is attacked. Bacteria can then reach the softer dentine below and make it decay. When the teeth are decayed, food cannot be chewed properly. Bacteria and poisons from the decayed teeth are swallowed with the food, and the person becomes ill.

However much care we take of our teeth we cannot always be quite sure that they are free from decay. It is foolish to allow teeth to decay so badly that they have to be pulled out. Decay can be arrested if it is in its early stages. The dentist will remove all the diseased part of the tooth, kill all the Bacteria that are there, and then fill up the hole with some solid substance. The dentist should be visited once or twice a year. He will then be able to detect the earliest beginnings of decay, and "fill" the tooth without pain. Teeth can be strengthened to resist decay in several ways. The jaws need exercise, that is hard foods should be eaten by children. Milk, butter, eggs and fresh green vegetables help to make the teeth healthy.

Cleanliness of the clothes and the surroundings

It is foolish to have a clean body and wear dirty clothes which will harbour the Bacteria which cause diseases. Clothes must be frequently cleaned.

No dust or dirt must be left anywhere in or about the home. Refuse left lying about forms breeding places for Bacteria, and Bacteria carriers, such as Flies, Fleas, Bugs and Lice, breed very quickly in dirty houses (see Chapter 8).

3. Exercise and rest

Exercise is necessary to strengthen all the muscles, to develop the brain and to keep the body in every way in a fit condition. Everyone should have as many exercises and games as possible in the open air.

Rest is as important to the development of a child's body as exercise. Children up to twelve years of age should have about twelve hours sleep each day. Adults require only eight hours sleep and sometimes even less. The body needs sound sleep if it is to have physical and mental health. The room in which one sleeps should be quiet, not too light, and furthermore it must be well ventilated.

4. *Wholesome food*

Good feeding plays a most important part in the health of the body. It is possible for a person to eat a normal amount of food, and yet the body may be starved. The body must have the proper amount of carbohydrates, fats, proteins, vitamins, salts of different kinds, and water. Milk is called a complete food because it contains all these different substances. Most other things that we eat contain some of these foods only, so it is important for us to have a mixed diet which will include enough of each kind of food. Vitamins are only required in small quantities, but they are so important that we must know a little about them.

Vitamin A is present in cod liver oil, milk, butter, cheese, egg-yolk, fresh green food, beef and mutton fat, suet, liver and carrots. It is most important for growth and helps the body to resist certain infectious diseases.

Vitamin B is present in cereals, pulses, yeast, milk, egg-yolk, liver, kidney, brains, cabbages, lettuce and watercress. It is necessary for growth and for the nervous system.

Vitamin C is found in lettuce, cabbages, oranges, tomatoes, lemons, potatoes, swedes, turnips and watercress. Vitamin C prevents scurvy. This disease was very common until some years ago in the navies and commercial ships. A person suffering from scurvy becomes very weak, and gradually becomes completely exhausted with constant pain in the muscles. Eventually scurvy may cause death.

Vitamin D is present in cod liver oil, oily fishes such as herrings, egg-yolk, milk, butter and animal fats. Vitamin D helps to develop and harden the bones and teeth. It prevents rickets, a disease very common in children.

Meals should be taken regularly, and as much fresh food as

possible should be eaten. Eating between meals is harmful and too many sweets should not be eaten.

5. *Protection of the body from harm*

A person who leads a healthy life is more resistant to disease (that is he is less likely to be affected by disease) than one who does not. Many diseases are caused by Bacteria which enter the body through the nose, mouth or skin. We should try to prevent Bacteria from entering the body, or if we are suffering from a disease ourselves we should do our utmost to prevent the Bacteria from spreading to other people.

Infectious diseases such as influenza and measles are caused by germs which we breathe in. A person suffering from an infectious disease should not cough or sneeze near someone else, because many germs will be in the drops of moisture which are forced out. A person is less likely to be infected in a well-ventilated room, as the infected air will be quickly removed.

The easiest way to prevent the spreading of an infectious disease is to put the infected person in a room by himself, that is, to *isolate* him. Unfortunately a person may have contracted a disease and may be infecting others for a short time before he feels ill. To prevent the spread of disease in this way, a person who has been in contact with an infected person is isolated until the disease has had time to appear. As infectious diseases are very common in schools we should learn a little about them.

Colds are infectious and can be passed from one person to another.

Influenza. The incubation period, or the time between the date of infection and the first signs of the disease, is about two days.

Signs of disease or symptoms: headache; pains in neck and limbs; dry cough; sometimes sickness and diarrhoea.

Measles. Incubation period is usually twelve days. The disease is most harmful to children from six months to five years old.

Symptoms: severe cold for four days. Then a rash appears on the face, which spreads to the body. It disappears after two days, then the body skins. Patients usually dislike strong light as their eyes are affected.

Measles is very harmful because it is often followed by diphtheria, pneumonia, or ear trouble.

A patient is infectious for three weeks after the beginning of the rash.

Scarlet Fever. Incubation period is two to four days.

Symptoms: headache, sore, red throat which swells and becomes painful; the tongue is coated; a rash appears on the second day on the neck and chest, and spreads to the trunk, arms and legs; a white patch appears round the mouth. The rash disappears after a few hours. The skin peels for several weeks.

The patient should be isolated for five to six weeks. Scarlet fever is not very harmful nowadays, but it may be followed by ear trouble.

Diphtheria may follow scarlet fever, measles or whooping cough. Incubation period is one to seven days.

Symptoms: the symptoms are similar to those of scarlet fever at first. Then the angles of the jaw swell and become tender. Yellowish-white spots appear on the tonsils, which spread and finally join to form a continuous membrane. A foul smell comes from the mouth and a discharge from the nose.

Before a person can come out of isolation, the germs must be missing for three consecutive weeks. Swabs of the throat are taken, that is a small piece of sterilised cotton wool is gently rubbed round the throat. The substance on the swab is then looked at under a microscope to see if any of the germs are present.

Heart disease or paralysis may follow diphtheria. This disease was, at one time, very dangerous but nowadays there are fewer deaths (see p. 138).

Chicken-Pox. The incubation period is ten to fifteen days. It occurs chiefly amongst children from two to six years old.

Symptoms: rash begins on body and then spreads to the face, scalp and limbs. The rash appears as red oval spots which change into small blisters or "pocks". These burst in one to two days, and a scab forms. The rash irritates very much, but the patient must not rub off the scabs or nasty scars will be left. The patient must be isolated until the scabs, which are very infectious, have fallen off.

Mumps. The incubation period is eighteen to twenty-two days.

Symptoms: swelling of the salivary glands, especially those immediately in front of the ears; pain in the ears and throat; sometimes vomiting and nose bleeding.

The patient is infectious one day before the swelling appears until one week after the swelling has gone. The patient should be isolated for three weeks.

Whooping Cough. The incubation period is usually about a fortnight, during which time a person is very infectious.

Symptoms: dry cough which gradually changes to the characteristic whoop.

The termination of the disease is difficult to determine as a patient may continue to whoop for some time.

With all these diseases the advice of a doctor should be asked for as soon as possible.

Immunity

If a person suffers from any of these diseases this is what happens. The germs, or Bacteria, in the blood multiply enormously and the patient is ill. Slowly the body makes chemical substances, called *antitoxins*. These either kill the Bacteria, or make them more tasty for the white corpuscles of the blood to eat.

When a person recovers from scarlet fever, measles or chicken-pox, he cannot catch that illness again for some time. He is said to be *immune*. This is because the body has put so much antitoxin into the blood, that if any of the Bacteria causing these diseases get into the blood, they are killed. A person is not rendered immune after an attack of all other infectious diseases, but he can be made immune. For small-pox, which is somewhat similar to chicken-pox, a person is *vaccinated*. A substance is put into the blood which gives the person a mild attack of a disease which resembles small-pox. The body then makes quantities of small-pox antitoxin. A person can also be made immune to diphtheria by injections.

Contagious diseases are those whose germs are carried from one person to another by actual contact. The spreading of these diseases can very easily be prevented if no one is allowed to touch anything belonging to the infected person. It is advisable to train children at an early age not to exchange hats or other articles of clothing. It is advisable always to use one's own towel, tooth brush, hair brush, etc.

Many skin diseases are contagious, impetigo (causing nasty-looking scabs on the face and scalp) and ringworm being quite common in some places. Small-pox, scarlet fever and measles are contagious as well as infectious.

Constipation

Before leaving the subject of how to protect the body from harm, we should say a little about constipation. Constipation is an unnatural condition which is very common amongst children and adults. It can cause much harm to the body and greatly affect the health of an individual. If the waste material in the intestine is not got rid of regularly, poisonous substances are absorbed into the blood. The organs of the body do not then function properly. The person becomes unhealthy and is not able to resist disease.

Constipation is not something which has to be cured, it should be prevented. Exercise is necessary for the proper functioning of the intestine. A person who is compelled to sit down all day at his work should have exercise during his leisure hours. Constipation may also be due to eating the wrong kind of food.

Natural foods, especially vegetable foodstuffs, contain a certain amount of indigestible material which must be got rid of. This acts as a stimulus to the intestine. If we eat food that is easily digested, the intestine is not stimulated naturally, the muscles in the wall of the intestine become slack and will not function properly. Artificial chemical aid has then to be sought in the form of pills which help to stimulate the muscles, but it is far better to take exercise and eat the right kinds of food so that pills are not necessary.

First aid in the home

There is not sufficient space here to describe fully all that a person should know about first aid. So we must just learn a little about first aid in the home. It is, however, advisable for everyone to acquaint themselves with such a book as the first aid handbook of the St John's Ambulance Association. Often, after an accident has occurred, proper treatment until the doctor arrives will avoid any further complications. Here are a few first aid hints that are useful in the home.

Cuts

If the skin is broken at all, care must be taken to prevent the entrance of germs. Even the smallest scratch may lead to serious consequences. Iodine put on to a small wound will kill all the germs. A larger wound must be thoroughly cleaned with water containing either a few boric acid crystals or a few drops of iodine. The water should have been previously boiled and allowed to cool. If a wound is very long and deep, the advice of a doctor should be sought, as it may be necessary to have it stitched. This will make the wound heal more quickly, and will prevent a nasty scar.

Bruises

These are caused by bleeding under the skin. A piece of lint soaked in equal parts of spirit and water, or ice or cold water dressings, should be applied.

Nose bleeding

1. Place the patient in a sitting position with the head slightly thrown back.

2. Apply cold water over the nose and also to the spine at the level of the collar. Place feet in hot water if possible.

3. Patient must breathe through the mouth and must not blow the nose in case the blood clot, which forms to stop the bleeding, is removed.

If the bleeding continues, the advice of a doctor should be sought.

Burns and scalds

Scalds are caused by moist heat such as hot water and steam. Burns are caused by dry heat, electricity, lightning, friction, an acid or an alkali such as caustic soda, ammonia or quicklime.

The effects of burns and scalds are reddening of the skin and blisters, or the skin may be burned black. The clothing may stick to the burned part. The chief dangers are shock and the entrance of Bacteria. The treatment necessary is as follows:

1. Carefully remove clothing from the injured part if you can do so easily.

2. Do not break the blisters.

3. Place the injured part in a soothing lotion until a suitable dressing can be applied. This will also soak off any clothing that may be sticking to the wound. A dessert spoonful of baking soda to a pint of warm water makes a good lotion.

4. Dress the wound with lint or gauze soaked in the baking soda lotion. If the injured part covers a large surface, it must be dressed in patches, as exposing a large surface may give the person a shock.

5. If the lotion cannot be made, cover the injured part with cotton wool or some soft material.

6. If a person is burned with an acid, wash the injured part quickly with baking soda lotion as above (washing soda could be used here instead). If this is not available wash the burnt part

with warm water to dilute the acid. Then apply treatment as for an ordinary burn.

7. Seek medical aid.

Fainting

A faint can often be prevented if a person is quickly placed in a chair, with knees widely separated. Thrust the head down as far as possible towards the floor.

If a person has fainted the following treatment should be applied:

1. The patient must be placed flat on the ground or on a bed where there is fresh air.

2. All tight clothing must be loosened.

3. Sprinkle the face with hot and cold water alternately, and apply warmth over the heart. Rub the limbs upwards.

4. Smelling salts may be held to the nose.

5. Stimulants such as brandy or whisky should not be given unless advised by the doctor. Tea, coffee, hot milk or water can be given.

Stings

There are several insects that may sting. The following treatment may be given:

1. Extract the sting if it is in the wound.

2. Mop the part with spirit, iodine or dilute ammonia. A solution of washing soda (one dessert spoonful to a pint of water) or rubbing the place with a wet blue bag will relieve the pain.

3. Do not make the sting worse by rubbing it much.

4. Put dry lint on the place.

5. If the sting is a bad one the advice of a doctor should be sought.

Poisons

It is advisable to see that all poisonous substances are carefully locked in a cupboard; then there is no fear of children taking them. Children, as well as adults of course, may be poisoned by tasting liquids in bottles left lying about, by eating food that is not good, or by eating poisonous berries or fungi. The rules for treatment are as follows:

1. Send for a doctor at once.

2. Keep any vessel that may have contained the poison, or any food or vomited matter.

3. If the lips and mouth are not burned, make the patient sick by giving an emetic every five minutes until vomiting occurs. Putting two fingers at the back of the throat sometimes hastens it.

Suitable emetics are made either by putting a tablespoonful of mustard or two tablespoonfuls of salt in half a pint of luke-warm water.

If the lips are burned it is best to give the patient a large amount of water to dilute the poison.

Olive or salad oil, medicinal paraffin, barley water and gruel all help to relieve the pain.

If a person who has been poisoned tries to go to sleep, he must be kept awake. This can be done by walking him about, slapping him with a wet towel and by giving him strong black coffee.

Sprains

It is very easy to sprain a joint by giving it a sudden twist. A sprained joint is very painful, and it may swell and become discoloured.

1. The limb must be placed in the most comfortable position.

2. Cold dressings should be applied. If these dressings do not give relief hot dressings should be applied instead.

3. Ask for medical advice.

You have just learned a few first aid hints that may be useful in the home. Everyone should have a First Aid Box in their house, so that when an accident occurs, they know exactly where to find all the necessary things that may be required.

Chapter 12

YOUNG ANIMALS

Animals and machines

Do you remember how a Frog begins life? It starts as a black **egg** surrounded by jelly. The eggs are laid by a female Frog in a pond, and each egg, before it can change into a Tadpole, must be *fertilised* by a **sperm** from a male Frog. You will remember that the Hydra starts life in just the same way (p. 65).

After it has been fertilised, the egg can be seen slowly to change into a Tadpole. At first the egg is just a little round ball of living matter, without any structure. Then gradually it gets a head, body and tail. Later on the Tadpole gets a mouth and eyes, and in the end legs and arms grow out and the Tadpole turns into a little Frog. Gradually organs are formed which were not there before.

This is the great difference between a living thing and a machine. An engine is made finished and complete. But an animal or a plant develops its parts gradually. A seed has very little structure, but slowly it turns into a complicated tree.

Development and growth

Until the Tadpole hatches from the egg, there is no growth. It is true that the egg has been changing, or *developing*, into the Tadpole, but it could not *grow* bigger because it could not eat. After it hatches, the Tadpole has a mouth, then it eats and can grow. These, then, are the two great changes during the time that an animal turns from a fertilised egg into a grown-up: it develops, and it grows.

After an animal has grown to its full size, it can produce a family of young, and then, after a time, it gets old, and finally it dies. This is the fate of all animals, unless they are killed by others before they grow old. But you will be very surprised to hear that parts of animals can go on living for ever, without growing old and dying.

Living cells that do not grow old or die

This is true, in the first place, of eggs and sperm. They can join together, when the egg is fertilised by the sperm, and then turn into a new animal. Later on the eggs or sperm of this new animal can themselves become yet another generation. And so on without end. Any animal alive to-day has come from an egg and a sperm. These were produced by animals perhaps now dead, and so on backwards in time. So, you see, the living cells of any animal, or human being, come from living cells far back through very many generations. So the living matter need not die: it can continue alive through eggs and sperm into the next, and the next, and the next generation, and so on.

That is how living matter may remain alive, without ever dying, through countless generations. But in addition to this, scientists can now keep parts of animals alive without their ever getting old. Little fragments of the muscle, the heart, the kidney, and other parts of Chickens have been kept alive for twenty-three years, which is much more than the whole life of a Fowl. All this long time the living cells continue to divide and multiply. They are given solutions of food for growth, and every other day the scientist must remove and throw away half of the newly formed cells, or they would become overcrowded. And through all the twenty-three years the living Chick cells have gone on growing and dividing without ever slackening or showing signs of age. This means that living matter need not get old: the Fowl must get old, but its cells, outside its body, need not get old or die.

Eggs

Now let us see how the Chicken develops in its egg. The egg of the Fowl is, of course, very much bigger than a Frog's egg. This is because it has much more *yolk* in it. The yolk is food

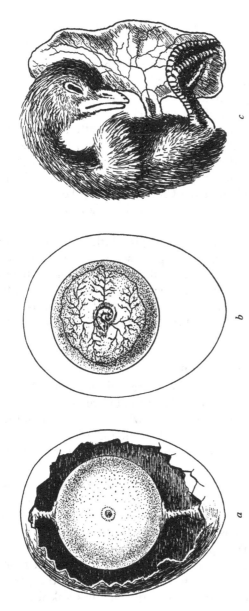

Fig. 118. Development of the Chick in the egg. *a*, An egg with the shell opened. The patch on the yolk will become the Chick. The yolk is slung in the white by two chords. *b*, The young Chick aged 4 days. The large head, the right eye, and the veins on the yolk can be seen. *c*, The Chick 19 days old, shortly before hatching. The wing and feathers are growing. The yolk is still attached to the navel.

for the Chick until it hatches. Since there is so much yolk, the Chicken develops a long way before it hatches. When a Tadpole hatches it has hardly got a head and it has no limbs yet, because the Frog's egg contains little yolk, but the Chick is a real Bird when it breaks out of the egg shell.

Since the Bird's egg is so well looked after, with its yolk for food, and so well protected by a shell, fewer eggs need be laid. A Frog lays hundreds, and most of the helpless Tadpoles perish. But a Hen, in nature, would only lay 8 or 10, and then get broody and look after the Chicks. We keep Hens in farmyards or gardens in a special way so that they will go on laying eggs.

The living part of a Hen's egg is a little transparent patch on one side of the yolk. If you get a setting of eggs and put them under a broody Hen, or keep them in an incubator, the Chicks will develop inside them. Then you can take eggs from time to time, open them carefully on one side, and look at what has been going on inside them (Fig. 118). There is nothing more interesting than this.

How the Chick develops

After a day you can see the future brain and spine of the Chick. A few hours later the little heart can be seen beating. It pumps red blood into arteries and veins which grow out on to the yolk. This blood brings yolk into the young Chick.

At the end of two days the beginnings of eyes and ears can be seen.

Before the third day is past the body gets enclosed in a transparent bag of liquid which acts as a water cushion.

After three days the wings and legs begin, but the young chick does not look in the least like a bird yet.

Then a beak appears, and after ten days feathers begin. By now the little arteries and veins have grown all round the yolk and they continue to bring in food from the yolk. Also they collect oxygen for breathing which has leaked in through the shell. These arteries and veins enter and leave the little Chick through its navel.

After three weeks the Chick pecks its way out of the egg shell. You see how the organs of the Chick appeared one after another, and gradually developed to their full form. They are not all ready-made like the parts of a machine.

Good and bad parents

A Chick has more yolk than a Tadpole while it is still in the egg. So it hatches later and is more developed. After hatching the Chick is protected by its mother for some time. But Frogs take no interest in their young. Even Sticklebacks are better parents (p. 87).

Some animals provide even less for their young than Frogs do. Starfishes, Corals, Sea-worms and Oysters have tiny eggs which you can only just see. They are so very small because they contain practically no yolk. Since there is hardly any food in these eggs, the young must hatch very quickly and are very little developed. They swim on the surface of the sea.

Because these young are so small and helpless when they hatch, enormous numbers of them perish. They are food for Fishes. Therefore vast numbers of eggs are laid, so that a few at least of the young will survive. A single Starfish lays several million eggs at a time.

Amongst the Insects we also find good and bad parents. Flies, Butterflies and Moths, for instance, lay large numbers of eggs which are unprotected. It is true that the mother lays her eggs close to the food which the young ones will eat when they hatch out of the eggs, but after laying the eggs she takes no further interest in her young.

Ichneumon Flies lay their eggs inside caterpillars. In this way the eggs are not only protected from harm, but when they hatch out their food is at hand. They eat the caterpillar itself.

Other Insects such as Hive Bees, Wasps and Ants, which are Social Insects, take care of their young until they are fully grown (Chapter 8). Solitary Bees and Wasps cannot tend their young in such a way. Instead they lay their eggs in small chambers which they make in the ground. Inside the chamber they put food for the young grub when it hatches out. The chamber is then sealed to prevent any harm coming to the young one. Fig. 119 shows a number of chambers of the Solitary Leaf-cutting Bee, made in a decaying beam of an old greenhouse. These chambers are made with leaves which the Bee has cut and fitted together.

Fig. 119. A number of leaf chambers of the Solitary Leaf-cutter Bee in a decaying beam of wood. In each chamber is a single egg with a store of honey and pollen for the food of the young when it hatches.

The young of Mammals

There are animals which look after their young even more than Birds do. These are the Mammals. The eggs of Mammals are very small indeed. They need not be big, because, after fertilisation by a sperm, the egg develops inside the body of the mother, where it is very well protected. The egg has no yolk, that is why it is so small, but the young animal into which it develops gets food from the blood of the mother. The food goes into the young animal through its navel. Inside the womb of the mother the organs of the young animal appear one after another and gradually develop, just as the organs of a Tadpole or a Chick do.

When the young Mammal is big enough, it is born. But even then the mother goes on feeding it with milk. And the parents teach the young how to find food and avoid enemies.

The young of Pouched Mammals like the Kangaroo are born very early indeed (they are only about an inch long), and then they are kept in the mother's pouch while they are fed on her milk.

Mammals look after their young so well, both before and after birth, that they only need to have few young; their families are very small compared with those of Frogs, Fishes, or Oysters.

Chapter 13

EVOLUTION

Life has existed on the earth for an immensely long time. The remains of animals and plants have been found which lived not thousands but millions of years ago. These remains are found in the rocks. They are called fossils. Most fossils are the hard parts of animals, the shells, skeletons and teeth.

Extinct animals

Most, but not all, of the numerous different kinds of animals whose remains are found in the rocks are no longer to be found alive on the earth. They are now extinct. The fossil bones of these extinct animals tell us that most of them were different from any animals living to-day.

Some animals are of course becoming extinct now, and in the course of the last few centuries animals once wild in England have disappeared. In the time of Henry VIII there were Wolves in this country. To-day there are none, although Wolves are still found wild in other parts of Europe.

But there were animals alive at the time when prehistoric man lived which are now quite extinct. A large kind of Elephant

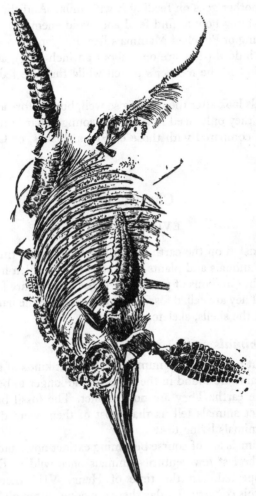

Fig. 120. Fossil skeletons of a reptile which lived in the sea and of its young. These animals lived 100 million years ago. Notice the paddle-like hands for swimming.

called the Mammoth was formerly very common all over northern Asia and Europe, including Britain. It had long coarse hair and huge curved-up tusks. Although Mammoths and prehistoric men lived side by side, Mammoths long ago died out. Yet complete carcasses of Mammoths, with skin, hair, trunk, muscles and bones, have been found frozen in Siberia. Most fossils, however, are the remains of animals and plants which died out long before the Mammoth.

The origin of fossils

How has it happened that the remains of extinct animals are found as fossils embedded in rocks?

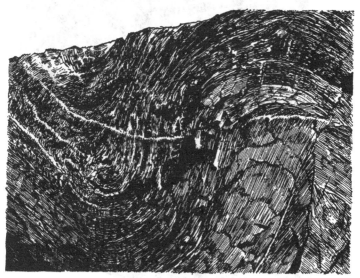

Fig. 121. Drawing of a cliff showing layers of rock formed from mud and sand at the mouth of a river. The layers have become bent by the earth movements that raised them above the sea level.

Rivers, flowing to the sea, carry with them immense quantities of mud in suspension. The Thames, for instance, brings down so much mud every year that if the mud were made into

Fig. 122. A fossil leaf of a Fern which lived in a Coal forest.
It was alive 300 million years ago.

stones it could build St Paul's Cathedral. At the mouths of the rivers this sediment is deposited on the sea bottom. The bigger particles of sand fall first. The finest mud is carried farther out to sea by currents before finally it falls on the sea bed.

When Fishes die, they fall to the bottom of the sea. If they are not chewed up by other sea creatures, the dead bodies will slowly be covered up by the falling mud. The same fate will happen to the carcasses of any land animals which have floated down with the rivers to the sea. In the mud the soft parts of the dead animals soon putrefy. They are destroyed by Bacteria. But the hard skeletons remain.

As layer upon layer of sediment is laid down, the mud or sand beneath is slowly turned into hard rock by the pressure of the layers above. The mud becomes clay, the sand is turned into sandstone. And slowly the bones or shells buried in the sediment are changed to fossils.

It is not only animal remains which can become fossils. Plants may also be preserved. Coal is formed from the remains of forests. Tree trunks and fossil leaves (Fig. 122) are found in the rocks which have layers of coal in them.

Earth movements

You see now how rocks are formed from sediment and how fossils may be shut up in these clays, sandstones, or chalk. But how is it that the rocks with their fossils are now found above the sea, even on mountain tops?

The reason is that the crust of the earth is not stationary. It moves up and down in different places. True, these movements are very slow indeed, so slow that they generally pass unnoticed. Yet there is clear evidence that such movements do take place. At many places on our coasts there are pebble sea beaches now well above the reach of the waves. Elsewhere on our coasts there are buried forests. Near Naples there is a ruined Roman temple. The lower parts of the columns are riddled with small holes. These were made by Mussels. Their shells are still in the holes.

This proves that after Roman times the temple sank below sea level. Now, once more, the land has risen, so that the temple ruins are again above the sea.

The temple near Naples has moved down and up again during the Christian era. But these two thousand years are nothing at all in comparison with the millions of years that animals and plants have been living on the earth. All that time slow movements of the earth's crust have gone on. Layers, or strata, of rocks laid down from sediment under the sea have been slowly raised up above the sea level. Many times in past ages different parts of the British Isles have been under the sea, and then raised up into land again. By such movements of the earth, mountain ranges are formed. Sometimes the movement is more violent. Then earthquakes result. In the raising and folding process the layers, originally flat, often become bent and curved, and you can see them like that in cliffs or quarries (Fig. 121).

Past ages

It is clear now that when a number of layers of rock are seen in a cliff or quarry, the upper layers of these rocks are younger than the lower layers. For the sediment was gradually laid down in the sea, one layer on top of the other. By comparing one cliff with another it has been possible to construct a complete section or diagram of all the strata on the earth's surface. Fig. 123 shows such a section. The top strata are the youngest, the bottom ones the oldest. Obviously the thickness of each kind of rock is roughly a measure of the time it took to be laid down in the sea. Actually we know that the Chalk was formed some 100,000,000 years ago. Coal dates 300,000,000 years back, and the lowest strata containing fossils are 600,000,000 years old.

During this immensely long series of ages the climate of each part of the world has changed a number of times. In Britain, for instance, the climate has sometimes been much warmer, sometimes much colder than it is at present. During the age when the

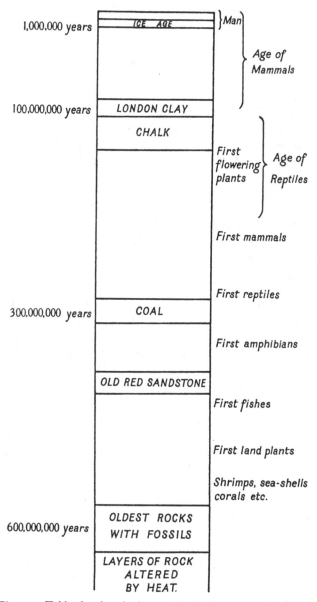

Fig. 123. Table showing the layers of rocks formed under the sea, with their ages and the chief fossils found in them.

Coal forests flourished the climate was warm. Later it became temperate again, but at the time when the clay which is beneath London, or London Clay (Fig. 123), was deposited, the sea was tropical. This is proved by the nature of the fossils. Coral reefs flourished in the sea which then covered part of England and we know that Corals only live in warm seas. Later on there was an Ice Age. The whole of northern Europe was covered by glaciers. In England they extended south as far as the Thames valley. Rocks scratched by this ice can be seen to-day.

Fossil Men

Man has only existed on the earth for a relatively short time, short, that is to say, in comparison with the periods of time we have spoken of. Human beings have been on the earth for only a million years. Fossil human skeletons have been dis-covered in a number of places and the stone tools and weapons which early prehistoric man made are found in many places. The strata in which these remains have been found prove that prehistoric man was on the earth before the Ice Age, but not very long before that time. Early human

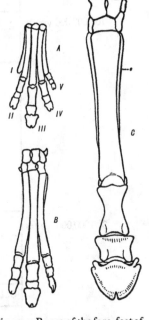

Fig. 124. Bones of the fore-feet of Horses. *A*, A fossil Horse which lived at the time the London Clay was formed; it had 3 toes on the ground (II, III, IV) and 2 toes not touching the ground (I and V). *B*, A larger fossil Horse, not so old; it only had 3 toes. *C*, A Horse of to-day; it is the largest and has only the middle toe and 2 useless splint bones (*c*).

beings differed from modern men in a number of ways. Their brains were smaller. In appearance they must have been some-thing like Apes.

Fossil Horses

As an example of what fossils can tell us about animals which existed in past ages we will take extinct Horses. You will remember that the hoof of the Horse is a huge nail on a single toe. On either side of the "cannon bone" of this middle or third toe there is a "splint bone" (Fig. 124 *C*). These splint bones are useless second and fourth toes. But at the time when the London Clay was deposited (Fig. 123) Horses had five toes, not one. Fig. 124 *A* shows the skeleton of the fore-foot of a Horse which lived a little after the time of the earliest five-toed Horse. You can see that its three middle toes touched the ground, but that the first and fifth did not do so. Fig. 124 *B* shows a Horse which lived in a yet later age, but still about 50,000,000 years ago. It walked on its middle toe only, but the second and fourth were still big. The first toe has disappeared, however, and the fifth is a mere splint bone. Fig. 124 *C* gives the foot of the modern Horse, with its single toe and cannon bone and its two splint bones.

The oldest Horses of this series were no bigger than Greyhounds. As we go upwards in the layers of rocks, each successive kind of Horse gets bigger in size, and at the same time its limbs become more and more suited to trotting and galloping.

It is not only extinct Horses that can be traced back into past time in this way. Other animals, Elephants and Camels for instance, have the same sort of past histories. As we go down in the strata of rocks, Elephants become gradually less and less like the modern Elephant. At the same time they have fewer and fewer peculiar features, such as trunks, tusks and crushing teeth. That is to say that the farther back we go in time, the more fossil Elephants were like other Mammals.

Extinct Reptiles

The oldest fossil Horses and Elephants are found at the time of the London Clay (Fig. 123). From that time onwards, Mammals were the most abundant large animals on the earth, just as they are to-day. But before the London Clay very few Mammals existed at all and they were very small.

Through long ages, however, before and during the time of the Chalk, Reptiles were very abundant (Fig. 123). Many of them were quite different from any Reptiles alive to-day. And many of these long-ago Reptiles were huge in size. They took the place on the earth later on occupied by Mammals. But just before the London Clay these giant Reptiles died out. They became extinct.

Fig. 125. What a giant extinct Reptile looked like that lived 100 million years ago. It was eighteen feet high and was a beast of prey.

Mammals to-day are some of them flesh-eaters, like Foxes and Tigers, others, such as Deer and Rabbits, are vegetable-feeders. In the past there were huge flesh-eating and vegetable-eating Reptiles. Among the Mammals of to-day, Whales and Seals live in the sea, while Bats fly. So, too, there were Reptiles ages ago which swam in the sea (Fig. 121) and others which flew (Fig. 126).

The first fossil skeleton of one of the huge land reptiles was found in Sussex. It fed on plants, and it stood on its hind legs like a Kangaroo. Its head was fourteen feet from the ground. There were also enormous flesh-eating Reptiles of much the same build (Fig. 125). The fact that they ate flesh while the others fed on plants is known from their teeth. Just as Lions prey on Antelopes to-day, so one kind of giant Reptile hunted another.

Huge as they were, these giant Reptiles are not the largest animals that have ever lived. The biggest Whales of to-day are the largest animals that have ever lived. Some of them are a hundred feet long.

Fig. 126. Two kinds of flying Reptiles as they must have appeared.

A most interesting fact about the giant Reptiles of long ago is that they all had very small brains. Perhaps one of the chief reasons why they died out was that they could not compete with the more cunning Mammals when these became abundant on the earth.

Flying Reptiles and fossil Birds

The most astonishing extinct Reptiles were those which flew. They are called Pterodactyls (Fig. 126). The wing was constructed on a different plan from those of Birds or of Bats (Fig. 89). The fourth finger was enormously long and a skin stretched from it to the side of the body and the leg (Fig. 126). Some Pterodactyls were of the size of Crows, others had wings eighteen feet across. The wings of a small aeroplane are only fifteen feet wide. But Pterodactyls can better be compared with gliders. The small size of the keel (p. 94) on their breast bone shows that their flying muscles were not big. It follows that the huge wings must have served only to glide and soar, making use of upward air currents like glider aeroplanes and some Birds such as Kites. They probably "took off" from cliffs.

Birds have existed on the earth from the Age of Reptiles (Fig. 123) onwards. No living Bird has teeth, but the earliest Birds had teeth like Reptiles. Moreover they had long tails. The tail was something like that of a Lizard with feathers on it.

Which animals came first on the earth?

Let us take the Animals with Backbones. Fishes first appeared on the earth at the time of the Old Red Sandstone (Fig. 123). Before that time there were no Animals with Backbones on the earth. Amphibians (p. 88) came later. They first put in an appearance just before the Coal forest Age. Soon after that time some of the Amphibians were very large compared with the Amphibians of to-day. They were something like Newts to look at, but were up to 9 ft. in length. Soon afterwards these giant Amphibians died out.

Then came the great Age of Reptiles (Fig. 123). Mammals and Birds appeared on the earth later than Reptiles. At first Mammals were not only few in number but they were small and insignificant. Then, at the time of the London Clay, Mammals of all sorts appeared. The giant Reptiles had died out by then and Mammals replaced them as the lords of creation. It was not until just before the Ice Age (Fig. 123) that Man arrived.

You will notice that the order in which the various classes of Animals with Backbones appeared on the earth is the same as the order in which we arranged them in Chapter 9: Fishes, Amphibians, Reptiles, Birds and Mammals, and finally Man. This is the order from the simplest to the most complicated kinds of Backboned Animals. In past ages, Amphibians, Reptiles and Mammals each flourished in turn as land dwellers. Mammals, with their controlled body temperature and better brains, had the last triumph, with Man at the end of the story.

Land plants tell the same story. The remains of Flowering Plants are found no lower than the Chalk Age. The Coal forests were formed by plants like giant Ferns and by other Flowerless Plants which afterwards died out. Flowering Plants are better for land life than Ferns, for they have seeds. So Flowering Plants succeeded, and gradually appeared in all varieties to take possession of the land.

The first animals

What animals inhabited the surface of the earth before the Fishes, the first Backboned Animals to appear? Below the first fossil Fishes found in the strata of rocks there are numerous remains of animals other than Backboned Animals. Before the Fishes there were already Corals, Whelks, Mussels, Shrimps and many others. Fossils of these are found right from the bottom to the top of the series of rocks. As with the Backboned Animals, the farther we go back in time, the less like were these creatures to animals alive to-day.

But it is quite sure that this planet was populated long before the time of the oldest rocks in which we find fossils (Fig. 123). Why is it then that we find no fossil remains of these very earliest animals?

The middle of the earth is composed largely of molten rock. Sometimes this comes to the surface in volcanoes. When this inner rock solidifies, it is crystalline. Granite is such a rock. Now the oldest strata of rock containing fossils have in past ages been forced down near to the hot central rock. The weight of strata above, and movements of the earth's crust, have been responsible for this. The result is that the lowest layers of rocks which once contained fossils have been literally cooked. The heat below and the pressure above have quite changed them. Slate is clay changed in this way. Limestone becomes marble.

But in this process all fossils are destroyed. The unfortunate result is that we can never know what the very oldest animals or plants were like. No fossils remain to tell us. We are sure only that there was life existing on the earth for a very long time before the oldest fossils. Probably the time from the beginning of life on the earth until the oldest fossils was quite as long as all the ages from then until to-day. That is to say, probably living things have been on the earth for a thousand million years.

Evolution

The numerous different breeds of Dogs are descended from a kind of wild Dog like a Wolf. In the course of time the various different breeds, Bulldogs, Dachshunds, Greyhounds and so on, have arisen. This is true not only of Dogs, but of all other domestic animals. And all the different kinds of wild animals and plants have a similar past history. To-day there are over a million different kinds of animals and plants on the earth. They are descended through the ages from other kinds of animals and plants. This process of descent is known as evolution. There is an immense amount of evidence for the truth of evolution.

Family histories

You will remember that as we go downwards through the strata of rocks, that is as we go backwards in time, fossil Horses get less and less like Horses alive to-day. At the same time the farther back we go, the less were fossil Horses different from other Mammals. Think of the gradual change from one-toed to five-toed feet as we pass downwards. The same is true for Elephants, Camels, and other animals and plants.

The questions are these: Are we looking at long family histories? Are modern Horses descended in a direct line from a five-toed animal? Or has each succeeding kind of Horse been created independently, lived for an age and then died out, to be replaced by a new creation? We believe that the first of these two alternatives is correct.

The different breeds of Dog all arose during the course of human history. The Dog breeds are all descended from a wild species of Dog, each breed becoming more and more unlike the wild Dog as time goes on. Horses, Asses and Zebras have a past history like Dogs. They are all descendants of the fossil Horses we have been talking about, from the small five-toed ancestors that lived 100,000,000 years ago. Not only is this true of Horses and Dogs. All living kinds of animals and plants are believed to be descended from other kinds now extinct.

All plants and animals alive to-day are twigs of a vast family tree. Present-day plants and animals are descended in direct lines from others now extinct and different from themselves. Some animals or plants have changed more, others less, in the course of time from what their ancestors were. Man and Apes, for example, certainly both belong to the same stock. Their bodies are very much alike. But in the course of evolution Man has changed more from the ancestor of both than Apes have done. There is definite evidence for this from fossil remains of Man, once called "missing links", now no longer missing.

The conquest of the land

The evolution of Amphibians, Reptiles, Birds and Mammals has been in the direction of better and better capacity for living on land. Amphibians are not so very different from Fishes. Not only are they still dependent on moisture when fully grown, but their young (Tadpoles) live in water. Reptiles, Birds and Mammals can live in quite dry air. It is Birds and Mammals which have gone farthest in evolution, with their constant body temperature and their superior brain power.

You have seen how the different kinds of Backboned Animals appeared one after the other on the earth. First came Fishes and then Amphibians. After the Amphibians, Reptiles appeared, then Mammals and Birds, and finally Man. But the family tree had many side branches which stopped long ago. Many lines of evolution went on for a time and then died out. The different kinds of huge Reptiles are good examples of this.

Life began in the water

Fossils show us how in turn Amphibians, Reptiles and Mammals, ending with Man, were masters of the land. Each of these kinds of Backboned Animals is in turn more and more suited to land life.

The ancestors of all the Backboned Animals were fish-like. They lived in water. So, too, the simplest animals of all, such as Amoeba, live in water. From some simple animal like Amoeba the ancestor of all animals evolved. This ancestor must have been something like Hydra, and it too was aquatic. Now, it is a fact that most groups of animals still live in water. Corals, Star-fishes, Oysters, Whelks, Shrimps, Crabs and Fishes are water-living. It is from aquatic ancestors that the land dwellers are descended. Three groups of animals only, namely the Insects, the Spiders and the Backboned Animals, have completely succeeded in conquering the land.

Plant evolution

On the plant side there is a similar story of gradual conquest of the land. The simplest plants, Bacteria, Yeasts, Pond Weeds and Sea Weeds, live in water. They have not evolved very far from the ancestors of all plants. Mosses are more fitted to land life and Ferns even better. It is the Flowering Plants, however, with their ways of avoiding water loss, and their seeds and pollen, which are most fully suited to land.

Now, the fossil history of plants tells how their evolution took place. After the lowest plants, Mosses appeared first on the earth. Then came Ferns, later on Pines, and lastly the Flowering Plants. Just as with the Backboned Animals, first the lower

(Flowerless) Plants and later on the higher (Flowering) Plants were in turn most abundant on land. And each group of plants had many side lines in its family tree, which flourished for a time and then died out to give way to more perfect land plants.

The beginning of life

What were the ancestors of all animals and plants like? What were the first living things to appear on the earth and start the family trees of animals and plants? We do not know.

There are two reasons why we cannot know this. Certainly the earliest living things must have been something like Bacteria or Amoeba. But they must have been too soft to become fossils. And, moreover, the oldest rocks formed of mud and·sand in the sea have been changed by heat and pressure (p. 162). Therefore all fossils in them have been destroyed. So no traces are left of any of the earliest plants or animals.

To-day, as far as we know, nothing living can come out of anything that is not alive. But at least once in the early history of the earth this must have happened—when life began.

The causes of evolution

There is no doubt of the fact of evolution. But for what reasons have plants and animals evolved? What has caused them in the course of ages to change from one kind into another?

As far back as the time of the ancient Greeks, scientific men suspected the existence of evolution. Later the idea disappeared, and right through the Middle Ages until the beginning of last century it was believed by everyone that all living animals and plants had been created a few thousand years ago. Fossils were still unknown and the real age of the earth was unsuspected.

In 1859 Charles Darwin published a famous book proving the truth of evolution. One of the reasons for the great success of Darwin's book was that it not only proved the reality of evolution but it also suggested a good reason why plants and animals have evolved.

Heredity

A child inherits features from its father and mother, and from its ancestors. There is a family resemblance between them. The same is true of animals and plants. Garden flowers and race horses are familiar examples. This is the fact of heredity.

Of late years a great deal has been discovered about heredity. It is possible to foretell, for example, when one parent has dark eyes and the other light-coloured eyes, what proportion of the children will have light eyes. A large number of facts of this sort have become known since the first discoveries of Mendel, an Austrian abbot, in the second part of last century.

Variation

Brothers and sisters are never exactly alike. Offspring of animals or plants always show differences among themselves. These differences are called variations. Brothers can vary in many ways. Their noses may be of different shapes. Their hair may vary in colour. They may be of varying heights. In the same way one brother may be more resistant to disease than another or he may learn more quickly.

Generally speaking, variations are due to one of two causes. Partly they are inborn. In other words, they are due to heredity: they are inherited. And partly variations are due to the life the individual has led. That is, they are caused by the surroundings.

Take the case of height. There are tall families of men and there are short families. On the average a member of a tall family will be taller than a member of a short family. But the exact height to which any member of a family grows will depend not only on his heredity. It will depend also on whether he was well fed or badly nourished as a boy. It may depend on whether or not he suffered serious illness. So too for plants. There are tall and short strains of Corn. But the exact height reached by any plant will depend also upon the amount of moisture and food salts in the soil. Both heredity and surroundings tell.

Occasionally quite *new* variations turn up in a family and are passed on by heredity to succeeding generations. Something which had never before been known in the family history appears. A strain of Rabbits may have been bred for generations having always the same coat colour. And then suddenly a few individuals turn up with differently coloured hair. Flowers may have been raised for generations and then unexpectedly a few plants appear with an unusual number of petals. Such new and unexpected variations are called ***mutations***. Once they have appeared they are passed on from parent to offspring by heredity. Up to now we know very little of the causes of mutations.

The struggle for existence

Many more plants and animals are born into the world than can possibly survive. You will remember how rapidly Bacteria multiply (p. 48). A single pair of Flies can produce 20,000 maggots. In a fortnight these turn into Flies and breed again. If all survived they would then give rise to 200,000,000 more larvae. A single Cod lays several million eggs. But of these, on the average, only two survive to grow into adult fishes. For neither Flies nor Cod are on the increase in numbers. So, on the average, only two offspring of a pair of parents can survive to replace their parents when the latter die.

This is true of all animals and plants, whether they have very numerous offspring like the Cod, or only a few. Most members of each family die young. They are either killed by enemies or they die for lack of food. Of all the seed scattered by a flower only very few can grow into new plants. Most are crowded out. There is no room for them in the soil.

But normally hard conditions of life and enemies keep down the numbers. There is a never-ending struggle for existence. Obviously in this struggle the weakest go to the wall. Individuals not vigorous enough to get most food go under. Those which are the strongest to fight foes or strongest to resist parasites live on and breed, while the weaker members of each family die young.

In other words, the result of the struggle for existence is a survival of the fittest.

Natural selection

Darwin saw that the struggle for existence should result in evolution. Those individuals which survive to breed must be the strongest, the healthiest, the most cunning. Now, we have seen that these characters of strength, health or cunning are due to two causes. In part they are family characteristics. That is to say, they are inherited. And in part they are due to outside influences, to the life the growing animal or plant has led. Variations, in other words, are due partly to heredity, partly to surroundings.

The result of natural selection is that those animals or plants survive which are best fitted to the life they have to lead. Now, when the survivors breed they will hand on to the next generation some of the characteristics which helped them to win in the struggle for existence. Those beneficial qualities which are hereditary will reappear in the next generation. In this way, generation by generation, any kind of plant or animal must get better and better fitted to its surroundings. Or, if no further improvement is possible, the high level of efficiency will be kept up.

The evolution of new kinds of animals and plants

So far you see how the survival of the fittest will maintain the efficiency of a kind of animal or plant. But how does it help to explain the evolution of one kind into other kinds in the course of time?

We must remember that the surroundings are not unchanging. In the course of millions of years there have been several cold periods and several tropical ages in northern Europe. We know indeed that climates change in much shorter periods of time. The Thames at London no longer freezes over as it once did. As these changes slowly occur plants and animals must change too. If they are to survive they must adapt themselves to the new conditions.

As the surroundings alter, natural selection will pick out new characteristics which previously had no value in the struggle for existence. Some characteristics, of course, will count for survival in the new climate which are due only to the action of the climate on the individual during its life. These characters are not passed on to the next generation, and so they will not count in evolution. But if an *inherited* character happens to fit the individual better to the new climate, then it too will be selected in the struggle. Individuals possessing the particular characteristic will survive and they will pass it on to their descendants. Thus gradually a certain kind of animal or plant will alter in many of its characters. It will turn into a new kind by the **natural selection of favourable inherited variations**. If you remember that from time to time *mutations*, that is to say *new* inherited variations, turn up, you will see that there must be ample material for natural selection to work upon and so produce new animals or plants.

Acquired characters

Of course, to a large extent animals and plants are able to fit themselves during their lifetime to changes in their surroundings. If a man becomes a blacksmith his arm muscles are developed more than those of other men. But will his sons, even if they are not blacksmiths, have very large arm muscles?

Now, if it were true that such characteristics *acquired* in the lifetime of a plant or animal could be inherited by its offspring, then this would supply a good explanation of evolution. As a climate gradually changed, for example, each succeeding generation would fit itself to the altered world (just as the blacksmith develops his arm muscles to fit his occupation) and would pass on to its successors what it has acquired. But it is unlikely that this ever happens.

Evolution seems to work by the more roundabout way of the

natural selection of mutations. For reasons we do not yet know, eggs or sperm are changed from time to time. This results in new characters, mutations, in the next generation. If the new characters are favourable in the struggle for existence, their possessors survive to pass them on to their offspring. Natural selection seems to be a sieve which separates good from bad mutations by killing off the bad.

Human evolution

This question as to whether or not characters acquired in the course of a lifetime are passed on by heredity is of immense importance, too, in considering human evolution. For with civilised Man, natural selection no longer has full play. The weakest are no longer allowed to go to the wall. The weakest have families just as well as the fittest. One of the few ways in which natural selection can work to-day is by the death of those less resistant to diseases.

But for generations now there have been schools, universities, trades. Does each generation pass on something of what it has learnt to its sons and daughters? Probably not. It is true that knowledge advances. But the advance accumulates in libraries. We are apparently no cleverer than our great-grandparents were.

Practical Work

N.B. Careful drawings should be made of all specimens examined. Carefully labelled diagrams are often more useful than notes.

Chapter 1

1. Examine the sleep movements of flowers such as the Daisy, Californian Poppy and Tulip.

2. Put a Clover plant into darkness. After an hour or so the leaves will have folded down.

3. Lightly stroke the lower side of a tendril of Bryony with a match. It gradually bends at the point touched.

Chapter 2

1. Examine a Buttercup flower and cut it through longitudinally. Use a lens.

2. Make a collection of wild flowers and name them. Wild flowers can be pressed, labelled and grouped according to
   (*a*) Families.
   (*b*) Habitats, i.e. where they grow.
   (*c*) The kind of soil in which they grow.

3. Look at pollen grains under the microscope. Shake pollen grains into 10 per cent. sugar solution and examine after a time to see the pollen tubes.

4. Study the structure of the Dead-nettle. Press down the "landing stage" and see what happens.

5. Make a collection of flowers similar to the Dandelion. Have they all two kinds of flowers? Look for flowers with ripe anthers and ripe stigmas.

6. Look at pin-eyed and thrum-eyed Primroses.

7. Examine the two types of Willow catkins. What is the difference between them?

8. Examine a Grass flower.

9. Examine the two kinds of Hazel flowers.

10. Make a collection of fruits. Group the fruits according to their method of dispersal. (Juicy fruits can be kept in a 2 per cent. solution of formaldehyde.)

11. Grow as many different kinds of seeds as you can. Note carefully the different stages in germination.

12. Find out under what conditions seeds grow best. Prepare a number of jars using Pea seeds only (see text p. 16). Put the seeds to grow under the following conditions:

Jar 1. With light, air, water and low temperature.
Jar 2. With light, air, water and high temperature.
Jar 3. With light, air, water and moderate temperature.
Jar 4. With light, air, no water and moderate temperature.
Jar 5. With light, water, no air and moderate temperature.
Jar 6. With air, water, no light and moderate temperature.
Jar 7. With light, air, water and moderate temperature using seeds killed by boiling them.

13. Grow a Bean plant in a pot. When the stem is about 5 inches long, put the pot sideways. Notice the curvatures.

14. Grow some Beans for a short time, then put them upside down and notice the curvatures.

15. Put a potted plant in a box with the light entering at one end only (Fig. 21). Note the result after a few days.

16. Set up the apparatus shown in Fig. 20. Note results.

17. Set Bean seeds at different depths in pots. Note when the shoots of each appear.

18. Get a clear plot of ground and set no seeds. Note what plants grow. Where have the seeds come from?

CHAPTER 3

1. Strip off the bark of a branch of a tree to see the green layer between the bark and the wood.

2. If you see the stump of a tree, find out how old the tree was by counting the annual rings.

3. Make carefully labelled drawings of (a) a Potato, (b) a Rhizome.

4. Cut a bulb in two and draw it, showing all the parts mentioned in the text.

5. Make lists of all the plants that you can find with underground stems. Head your lists (a) Runners, (b) Rhizomes, (c) Tubers, (d) Bulbs.

6. Set Carrot or Parsnip seeds and watch them grow. Compare the tap root after the first year's growth, and after the second year. What has happened to the tap root and why?

7. Put Willow or Poplar cuttings in water. Watch the growth of adventitious roots.

8. Collect stems of different ages and compare them.

9. Make a collection of the bark of trees, and name your specimens.

10. Find as many climbing plants as you can, and notice how they climb.

11. Germinate Pea or Bean seeds and draw the roots at different stages to show the parts of roots.

12. Examine the roots of plants in gardens. Make drawings and describe the following roots: Carrot, Turnip, Wheat, Cabbage, Lettuce, Pea and Grass.

13. Experiments to show that plants will only take in substances in solution.

Take two jars filled with water. To one add red ink, and to the other powdered carmine which will not dissolve in the water. Place a small plant in each with its roots dipping in the liquid. Leave for several days, then cut the roots above the surface of the liquid. Why is one coloured and not the other? Which one is coloured?

14. Examine the plants in your garden with woody stems. Are they annuals or do they live throughout the winter?

CHAPTER 4

1. Cut a Brussel Sprout (which is a large bud) in two lengthwise. Draw it to show its structure.

2. Put the twigs of various trees in water in a warm room. Draw them at different stages as they open.

3. Try to find a forked branch of a Horsechestnut or Sycamore tree. They will be at the ends of the branches where the flowers once grew.

4. Make a collection of twigs of common trees. Draw them and notice particularly the size, shape, colour and arrangement of the buds, the colour of the stem, the leaf scars, breathing holes and growth rings. Learn to recognise trees by their twigs.

5. Make a collection of leaves. Draw some of them to show (1) Leaves with and without stalks. (2) The different arrangements of the veins. (3) The various shapes of leaves. (4) The extent to which the blade may be cut up. (5) Simple and compound leaves.

6. Make a collection of climbing plants. Draw examples of plants that climb by: (1) Twining stems. (2) Thorns. (3) Tendrils.

7. Strip the skin off an Iris plant. Examine it under a microscope and draw it to show the pores.

8. Experiment to show that roots take in water.

Put a small plant through the cork in a test tube full of water (Fig. 48). Leave it for several hours and note the level of the water. The difference in the height of the water shows the amount of water taken in by the roots.

9. To show that leaves give out water.

Take a wide-necked jar which has a cork. Fill it with water. Split the cork and put a leafy shoot through it into the water. Make the cork air-tight with wax so that no water can evaporate from the jar. Cover this apparatus with a bell jar. Set up a similar apparatus omitting the leaves, and leave for a day or two (Fig. 127).

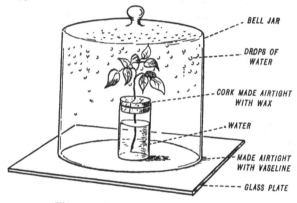

Fig. 127. Apparatus for Experiment 9.

Drops of water will be seen on the inside of the bell jar containing leaves, but not on the other bell jar. This shows that the leaves have given out water.

9*A*. Experiment 9 can be repeated using Laurel leaves that have been smeared on (*a*) the upper, (*b*) the lower, (*c*) both surfaces with vaseline. A comparison can be made of the amount of water given out in each case.

10. A rough comparison of the amount of water transpired by the upper and lower surfaces can be made in the following way.

Dissolve crystals of cobalt chloride in water. Dip pieces of white blotting paper into the liquid and dry the paper. The paper is pink when wet and blue when dry. Take a potted plant, cover the upper and lower surfaces of a leaf with cobalt chloride paper then a piece of glass, to keep out the air, and clamp it together. Support the leaf so as to keep it in its normal position.

As the leaf transpires the paper will become pink, so it is easy to see which surface gives out the most water.

11. The rate of transpiration under varying conditions can be told by using a Potometer, as shown in Fig. 128.

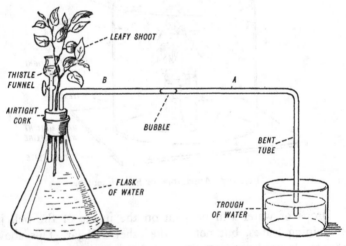

Fig. 128. A Potometer. See Experiment 11.

Take a flask fitted with a three-holed rubber cork. Through one hole push the cut end of a leafy shoot, through the second put a funnel fitted with a stopcock, and through the third put a tube bent at right angles. Fill the flask completely with water, insert the cork and make the joints air tight with wax. Immediately on filling, place the end of the bent tube under water. The rate of transpiration can be compared by admitting a bubble of air into the tube, by lifting the end of the bent tube out of the water in the trough, when the stopcock is closed. The rate of transpiration can be told by timing how long the bubble takes to get from A to B, because as the leaves give out water, more water is drawn along the tube from the trough.

The apparatus should be put into places where it is (a) sunny, (b) shady, (c) dark, (d) windy, to show the varying rates under different conditions.

The bubble can be made to return to A by opening the stopcock in the funnel, so letting more water into the flask.

12. Stir starch and sugar in cold water to show that starch will not dissolve but sugar will.

13. Put a drop of iodine on starch. It turns dark blue immediately.

14. To test for starch in a leaf.

Put the leaf in boiling water for a short time to kill it. Soak the leaf in alcohol to get rid of the chlorophyll. Then put the leaf into a weak solution of iodine. If starch is present, the leaf becomes dark blue.

15. Experiment to show that carbon dioxide is necessary for the formation of starch.

Take a leafy shoot that has been in the dark for several hours, so that it contains no starch. Place the shoot in a jar of water under a bell jar, together with a dish containing caustic soda (which absorbs carbon dioxide). Stand the bell jar on a sheet of glass, then it can be sealed down with vaseline. Set up a similar apparatus omitting the caustic soda. After several hours test the leaves of both plants for starch. The leaves which had no supply of carbon dioxide contain no starch.

16. Fig. 129 shows the apparatus that can be used to show that leaves give out oxygen.

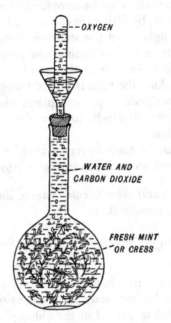

OXYGEN

WATER AND
CARBON DIOXIDE

FRESH MINT
OR CRESS

Fig. 129.

CHAPTER 5

1. Obtain Yeast from a bakery. Examine it in water under the microscope and draw it.

2. The following simple experiment can be done to show that Yeast forms a gas when it ferments sugar.

Put a little Yeast into a test tube containing a solution of sugar. Put a small bladder on the end of the tube. After a short time the bladder gradually swells up.

3. To show that Yeast forms carbon dioxide with malt.
Get a 2 per cent. solution of malt from a chemist. Put this solution and a little Yeast into a test tube with a bored cork, with a tube leading into a vessel containing lime water. The lime water gradually becomes "milky", showing that carbon dioxide has been formed.

4. Look at Mucor with a lens. Notice the sporangia.

5. Carefully look in the soil surrounding a Mushroom, or any similar Fungus that grows in the soil, for the white threads.

6. Draw a Mushroom to show the gills.

7. Shake the head of a ripe Mushroom on to a piece of white paper to see the spores.

8. Make a collection of Fungi. With the help of the reference books given, try to recognise them.

9. Make a collection of Mosses. Notice where different Mosses grow. Look for the capsules.

10. Draw a Fern plant and name the parts.

11. Examine the brown patches on the back of a Fern leaf (see text, p. 55).

CHAPTER 6

1. Put equal quantities of clay and sand into two jars of water and stir. Which settles first and why?

2. Germinate seeds in clay, sand and humus, and note the results.

3. Test various soils for water-retaining capacity (see p. 58).

4. Grow seeds in water-logged soil and well-drained soil. Note results. (Two flower pots one with and one without a hole can be used.)

5. To find the pore space of soil.

Put a known quantity of soil into a measuring jar and add a known quantity of water. Note the reading. Subtract the reading from the sum of the quantities of soil + water to get the pore space.

$$
\begin{aligned}
\text{Quantity of soil} &= \qquad + \\
\text{Quantity of water} &= \\
\text{Total} &= \qquad - \\
\text{Reading} &= \\
\text{Pore space} &=
\end{aligned}
$$

6. Make lists of the names of all plants found on several different patches of ground in different places. At the same time find out what kind of soil is there and compare your results.

7. Write down the names of any plants that you find in water-logged soil.

8. Experiment to find the amount of humus in soil.

Take a weighed amount of dried soil and put it in a tin over a flame. Heat it strongly for 15 min. Allow to cool then weigh the soil again. The difference in weight is the weight of the humus.

CHAPTER 7

1. Test any animals that you have in your school to find whether they can see, hear, smell, taste or feel.

2. Strip off the skin of a leaf, for example, Iris, and look at it under the microscope to see the cells.

3. Scrape the inside of your cheek, and examine the cells so obtained under the microscope.

4. Study living Amoeba under a microscope.

5. Examine living Hydra with a lens. Note retraction when it is touched and expansion again after a time. Look for buds.

6. Cut off the tentacles of Hydra. In a few days new ones will have grown.

7. For demonstration living sperm cells can be taken from the genital duct of a Water Snail and examined under a microscope in the blood of the Snail.

8. Examine with a lens an Earthworm which has been killed by putting it into boiling water.

9. Study the movements of an Earthworm as it moves forwards. Touch the front end of the Earthworm and see what happens.

10. Dip a glass rod in xylol and put it near the Worm. It will move away. Put the rod near different parts, and find which part of the Worm is most sensitive to smell.

11. Put some Worms in a jar of moist soil, with leaves placed on top. Look for the leaves in a day or two.

12. Examine the external features of a Crayfish or Lobster.

13. Keep a Spider in a jar containing twigs. Examine it carefully. Watch it spin a web. Feed the Spider with living flies and watch what the Spider does.

14. Find as many different kinds of webs as you can. Cut the "telegraph wire" of one web. Then put a live fly in the web. Does the Spider come to the web?

15. Put a Water Snail in a jam jar. Notice the waves passing forward over the surface of the foot as it moves; also its mouth opening and closing showing its tongue. Look for eyes and breathing pore.

16. Take a Land Snail and show that it can smell by putting a glass rod dipped in xylol near to the tentacles, when they are withdrawn. Move a pencil towards its eyes to find out how well a Snail can see.

17. Put an Earthworm into an aquarium containing a Leech and watch how the Leech feeds.

CHAPTER 8

1. Examine, with a lens, a Cockroach or any other large insect.

2. Collect as many different kinds of insects as possible. Keep them alive and study their habits.

3. Examine a Grasshopper. Look with a lens for the sound-producing organs and the ear drums.

4. Find and keep alive any stage of Butterfly or Moth and note their behaviour, date of hatching of eggs, moults and pupation.

5. Watch Butterflies feeding on a drop of sugar solution.

6. Examine living larvae of House fly or Blow fly. Notice how they crawl away from the light. Examine the pupae and adult flies.

7. Examine male and female Gnats.

8. Catch Gnat larvae and pupae and study their behaviour and structure. Cover the surface of the water with oil or petrol and see what happens.

9. Make a collection of insects which are harmful to plants.

10. Watch Bees collecting nectar and pollen and pollinating flowers.

11. Make a note of all insects that you see pollinating flowers.

CHAPTER 9

1. Draw a Herring to show all the external parts mentioned in this book. Lift up the gill cover to see the gills.

2. Watch a Fish swimming. Notice the movement of the fins, and the opening and closing of the mouth and gill covers when breathing.

3. Examine a Plaice or a Sole and notice the peculiar shape of its head.

4. Get some Frog spawn and keep it in an aquarium. Draw the different stages and note the date when each stage is reached.

[It is useful to keep preserved specimens of different stages. This can be done by putting specimens into a 2 per cent. solution of formaldehyde.]

5. When the Tadpoles are nearly fully grown, before their legs appear, feed them on meat. If small pieces of meat are tied with cotton, they can be suspended in the water for a short time and then removed. You will see the Tadpoles clinging to the meat, and eating it.

6. Examine the tongue of a Frog. It is fastened at the front end of its mouth and not at the back as our tongue is.

7. Make a collection of Birds' eggs and nests, and name your specimens. [Eggs should not be taken out of nests unless they have been forsaken by the parent Birds, or one egg can be taken out of a nest containing several eggs. Nests should not be collected until after the breeding season.]

8. Make a list of the Birds you see that (a) hop, (b) walk.

9. Write down the names of all the Birds you see in summer and in winter.

10. Examine the paws of a Dog and a Cat. How do they differ? Why?

11. Look closely at any Hoofed Animals you can and say whether they are odd-toed or even-toed.

12. If it is possible look at a Cow's stomach to see the structure. [Cows' stomachs are difficult to obtain as they are sold as "tripe" which can be eaten.]

13. Draw a feather. Look at it under a microscope or lens to see the barbules.

14. Collect pictures of Animals with Backbones, and put them into the groups mentioned in this book.

15. It is very difficult and in some cases impossible to study the animals mentioned in this book in their natural surroundings. Several animals can be kept in captivity with little trouble. These may be caught, or bought from a live-stock dealer. Visits could be paid to zoological gardens.

16. Visit any museums that are near to see specimens of animals.

CHAPTER 10

1. Examine the skull of a Rabbit, Cat or Dog to see how the bones are joined together.

2. If possible examine a human skeleton. If not try to obtain the skeleton of any other Mammal and compare it with the human skeleton.

3. Look at the shoulder blade of a Sheep.

4. Examine the teeth of a Cat or Dog. Compare them with your own.

5. Mix a little starch with cold water. Add a small amount of iodine. A blue colour appears. Add a little of your own saliva to the liquid, and the blue colour disappears as the starch is changed to sugar.

6. Obtain a Sheep's heart from a butcher. Examine its structure.

The action of the valves between the auricles and ventricles can easily be shown in the following way. Tie the arteries which lead away from the ventricles. Cut away most of the auricles, then pour water into the ventricles. Gently squeeze the ventricles when they are nearly filled, and you will see the valves come together.

7. Examine blood under a microscope.

Twist a piece of string tightly round the end joint of the middle finger of the left hand. Prick the tip with a sterilised needle to obtain a drop of blood. Put the blood on a glass slide, and cover it with a cover-slip to spread it out evenly. Then look at it under a microscope.

CHAPTER 12

1. Examine the developing chicks inside eggs taken at intervals from under a broody hen or from an incubator as described in Chapter 12.

CHAPTER 13

1. If there are any fossils in quarries or cliffs near your school, look for them and carefully chip them out with a cold chisel and hammer.

2. If you can do so, go to a museum and look at fossils.

Questions

When possible, the answers should be illustrated by means of clearly labelled diagrams.

CHAPTER 1

1. Make a list of the living and non-living things around you.

2. How can you tell living from non-living things?

3. Is it true to say that animals move from place to place, and plants do not? Give reasons for your answer.

4. "Plants make the whole world's food supply." Is this statement true? Why?

5. Give three examples of movement in plants. What are the uses of these movements to the plant?

CHAPTER 2

1. Which is the better method, Insect pollination or Wind pollination? Why?

2. Is it of any advantage to plants to have their pistils and stamens in different flowers? Give your reasons, with examples.

3. What are the chief features of (a) Wind, (b) Insect pollinated flowers? Give an example of each.

4. Why must seeds be scattered? Briefly describe, giving examples, the four ways by which seeds are dispersed.

5. What is meant by Germination? What conditions must be fulfilled before seeds will germinate?

CHAPTER 3

1. Plants usually reproduce by means of seeds. In what other ways can plants reproduce?

2. Why do annuals have slender green stems and perennials thick, brown stems?

3. How can you tell the age of a tree? What is your reason for saying so?

4. On which part of a root do the root hairs grow? Of what use are the root hairs?

5. Why is the root tip covered with a cap?

6. What is the difference between a tap root and a fibrous root? Of what use is each kind of root to the plant?

7. What is an adventitious root? Give three examples of adventitious roots.

8. What are the uses of (a) roots, (b) stems?

9. What is (a) an annual, (b) a biennial, (c) a perennial? Give examples of each kind of plant.

10. How does a stem increase in thickness? Why is there a difference between the wood formed in the spring and that formed in the autumn?

11. What is (*a*) sap wood, (*b*) heart wood? Of what use to the plant is each kind of wood?

12. How is a knot in wood formed?

13. Describe the life of a Carrot from the time the seed is set until the plant dies.

CHAPTER 4

1. Describe the various ways in which buds are protected. Why are they protected?

2. How are the "growth rings" on twigs formed? Why are they so called?

3. With the help of drawings, describe the structure of a leaf. Of what use are the pores?

4. The leaves of a plant are arranged so that they do not overlap. Say how this is done. Why are they so arranged?

5. Why is it important for leaves to transpire? What conditions affect the amount of water transpired by the leaves?

6. Explain exactly what happens before a leaf falls off a tree. Why do some trees lose their leaves during the autumn?

7. How do plants obtain their food?

8. What is meant by photosynthesis? Explain what happens in the leaf during photosynthesis.

9. What conditions are necessary before photosynthesis can take place? How could you show that your answer is correct?

10. The food stored in a Potato is starch. Where has the starch come from?

11. "Animals depend entirely on plants." Is this true? Give reasons for your answer.

12. What is meant by respiration? When do plants respire and why?

CHAPTER 5

1. By means of drawings, show how the Yeast plant reproduces. Compare this with the method of reproduction in the Amoeba.

2. Write a short description of the uses of Yeast.

3. Say all that you can about harmful and useful Bacteria.

4. How can you prevent food from being attacked by Bacteria?

5. Write a short essay on harmful Fungi, and say why they are harmful.

6. Describe a Moss plant. Compare it with any common flowering plant such as a Buttercup.

7. Describe a Fern plant as it would appear if you looked at it (*a*) during the winter, (*b*) during the summer.

CHAPTER 6

1. How does sand differ from clay? Illustrate your answer by describing any experiments that you know.

2. If there was too much clay in your garden soil, how could you improve it?

3. When a soil is water-logged, how can a farmer make it fit for growing crops?

4. After it has been raining, why do you see puddles on some soils and not on others?

5. What is meant by loam? Is this a good kind of soil for plants? Give reasons.

6. How does water help in the formation of soil?

7. Write a short essay on how soil is formed.

8. Why is lime added to a clay soil? Illustrate your answer with an experiment.

9. Why is it not advisable to grow the same crop on a piece of land for several successive years?

10. What conditions of the soil affect the growth of plants?

11. Why are both sand and clay bad soils in which to grow plants? What is the best kind of soil for plants?

CHAPTER 7

1. Do you think the Amoeba is a wonderful animal? Why?

2. On what does a Hydra feed? How does it catch its food?

3. Describe the ways by which a Hydra can reproduce itself.

4. How is a Coral formed?

5. Of what use are Earthworms in the garden?

6. Describe how Spiders catch their food.

7. Write a short essay on Snails and animals similar to the Snail. Say which are useful to man and which harmful and why.

CHAPTER 8

1. How would you recognise an insect? Make lists of the insects you know (a) with two pairs of wings, (b) with one pair of wings, (c) with no wings.

2. How do the Amoeba, Earthworm and Butterfly breathe?

3. How can you tell a female from a male Gnat? Which is dangerous? Why?

4. Give two examples of insects which spread disease. What is the special danger of the House fly?

5. Name six insects useful to man and six that are harmful. Give reasons for putting them into each group.

6. What insects that you know make noises? Describe how these noises are made.

7. What do we mean by "Social Insects"? Describe the life in a bee hive.

CHAPTER 9

1. Name three Fishes that you could catch in a river, and three in the sea. What two Fishes live in both fresh and salt water?

2. How does a Fish breathe?

3. Compare the life histories of the Eel and the Salmon.

4. What is peculiar about a Flat Fish?

5. Why do many Fishes lay a large number of eggs?

6. How do (a) Sticklebacks, (b) Newts protect their eggs?

7. What are the chief differences between the development of a Frog and that of a Newt?

8. Write a short essay on English Snakes.

9. Why is the bite of most Snakes poisonous?

10. How can a Snake swallow its prey whole, even when the prey is bigger than its mouth?

11. Where do Crocodiles live? Why are they dangerous animals?

12. Of what use is a Bird's beak? By means of drawings describe three different kinds of beaks and say what its particular use is.

13. How can Birds keep a store of oxygen?

14. Write a short essay on Birds' nests saying where nests can be found, and what materials are used in making nests. Describe the nests of any two Birds.

15. What are the chief differences between Reptiles, Birds and Mammals?

16. How is a Bird adapted to flying?

17. Compare and contrast a Bat and a Bird.

18. What is the difference between Whalebone Whales and Toothed Whales? Of what value to us are Whales?

19. What do we mean by a "warm-blooded" animal? What groups of animals are warm-blooded?

20. Name any animals that hibernate. Why do they hibernate?

21. What is a Mammal? Name one animal in each group of Mammals mentioned in this book and say why you put it in the group.

22. Write a short description of a Bat.

23. What animals are called Beasts of Prey? Why are they so called?

24. Compare and contrast Rabbits and Hares. How are the young protected?

25. Find out all you can about a Hedgehog, and write a short description of it.

26. What are the chief differences between an Indian and an African Elephant? Of what use is each kind of Elephant to Man?

27. How could you explain that a Whale is a Mammal and not a Fish?

CHAPTER 10

1. What is the skeleton of the body? What are the uses of the skeleton?

2. How are the brain, spinal cord, heart and lungs protected?

3. Compare the structure of a leg and an arm.

4. Describe the action of the biceps muscle.

5. Describe the course taken by food and drink after it is put into the mouth.

6. What is meant by digestion?

7. Write a short essay on teeth. Describe the structure of a tooth. How can we keep our teeth clean and healthy?

8. Give a brief description of the liver, and say of what use it is.

9. What is the course taken by air during breathing (*a*) with the mouth open, (*b*) with the mouth closed? Why should you breathe with your mouth shut?

10. Give a brief description of what happens to the blood (*a*) in the lungs, (*b*) in the capillaries.

11. Of what does blood consist? What are the uses of the blood?

12. In what part of the body are the kidneys? What work do they do?

13. With the help of a diagram, describe the structure of the skin.

14. Give a brief description of the different ways that the body gets rid of waste material.

15. How do the walls of the thorax move during respiration?

16. What is meant by (*a*) warm-blooded, (*b*) cold-blooded animals? Give examples of each.
How is the temperature of our body controlled?

17. What is meant by a reflex action? Illustrate your answer with an example.

CHAPTER 11

1. What things are essential if we wish to have good health?

2. Write a short essay on "Where there's dirt there's danger".

3. How do "germs" enter the body? How does the body try to get rid of these "germs"?

4. What do we mean by immunity? How can we become immune to a disease?

5. How could you tell that a child had (a) Measles, (b) Chicken-pox, (c) Whooping cough, (d) Mumps?

6. Why is it so important to take care of the slightest scratch that you may get on the skin? What is the best method of cleaning a wound?

7. If a person's nose was bleeding very badly, what would you do?

8. How would you treat a person who had been burned by (a) a fire, (b) steam, (c) an acid?

9. What are the general rules for the treatment of poisoning?

10. How can the spread of infectious diseases be prevented?

11. What are the effects of constipation? How can constipation be prevented? Why is it dangerous?

CHAPTER 12

1. Describe the development of a Chicken.

2. Why is it that Birds and Mammals need not have such large families as Frogs and Fishes?

3. Write a short essay on "Care of the Young amongst Animals".

CHAPTER 13

1. What are Fossils? How were they formed?

2. How can we tell the age of Fossils?

3. Briefly say anything you can about the evolution of the Horse.

4. What do we mean when we say that we are living in the Age of Mammals? What kind of animals populated this country before the Mammals appeared? Say anything that you can about these animals.

5. In some rocks we do not find Fossils. Does this mean that no animals or plants existed when these rocks were formed? Give reasons for your answer.

6. Why is it that brothers and sisters vary very much in spite of the fact that they have the same parents?

7. Although so many animals and plants produce large numbers of offspring, why is it that the world does not become over-populated by them?

8. What do we mean by survival of the fittest? Is this true amongst human beings? Why?

9. How do new kinds of animals and plants evolve?

Appendix A

QUESTIONS REQUIRING SHORT ANSWERS

These are intended for periodic test purposes. In the majority of cases the answers should not occupy more than one line.

CHAPTER 1

1. What is meant by Reproduction?

2. What do plants feed on?

CHAPTER 2

1. What are the special functions of a flower?

2. What is cross-pollination? What is its advantage over self-pollination?

3. Why do insects visit flowers? How do flowers benefit by their visits?

4. Of what use to a growing seed is the food contained in it?

5. What is the difference between a seed and a fruit?

CHAPTER 3

1. Of what use are the "eyes" of a Potato?

2. Why is the outer skin of a perennial tough?

3. What is the use of the runners of the Strawberry plant?

4. Why does the bark of trees crack or peel off?

5. How can you tell a tree from a shrub?

6. What are annual rings?

7. What colour is a Potato that has not been covered with earth? Why?

CHAPTER 4

1. How can you make a hedge grow thickly?

2. How could you tell a Sycamore from an Ash twig?

3. What is a "dwarf twig"? On what tree are they often found?

4. What is (a) a simple, (b) a compound leaf? Give two examples of each.

5. Of what use are the veins in a leaf?

6. What is meant by transpiration?

7. Why do cut plants wither quickly in a warm atmosphere?

8. Plants living in dry places often have small leaves. Why is this so?

9. What is meant by (a) deciduous, (b) evergreen plants?

10. How do water plants obtain their oxygen?

11. Why do the roots in a water-logged soil die?

12. Should plants be taken out of a sickroom at night? Why? What kind of plants should be allowed in a sickroom during the daytime?

CHAPTER 5

1. Give six examples of flowerless plants and say where they are found.

2. Where can Yeast be found? Do bakers obtain their yeast from this source?

3. Of what use are the spores of Bacteria?

4. If a jar of jam is attacked by Mould, why is it that the mould does not spread to the bottom of the jar?

5. What is the green substance in gorgonzola cheese?

6. What Fungi are useful to us and of what use are they?

7. What are the brown patches on the back of a Fern leaf?

CHAPTER 6

1. Why are boulders along the banks of rivers, or by the sea-shore, rounded?

2. What is meant by the "pore space" of soil?

3. What is peat? What use can we make of it?

4. Name the three chief constituents of soil.

5. What is meant by the "Rotation of Crops"?

6. What agents help to break up rocks to form soil?

CHAPTER 7

1. Into what two large groups can all animals be placed?

2. What diseases are caused by some Amoeba-like animals?

3. What happens to a Worm if it is cut in two?

4. How do Earthworms make their burrows?

5. Name all places where animals similar to the Worm can be found.

6. What use was made of Leeches some years ago?

7. What colour are Shrimps, Lobsters, and Crabs when alive, and when dead?

8. What do Snails eat? How do they eat their food?

9. Find out the difference between Snails and Slugs.

CHAPTER 8

1. Why is a Spider not called an insect?

2. What insect is the chief enemy of the Cabbage Butterfly? Why is it?

3. Most insects, and also other animals, multiply very quickly. How is it that we are not overrun by them?

4. How can you tell Butterflies from Moths?

5. How can we prevent House flies from spreading disease?

6. How was the Black Death caused?

7. What is "Blight"? How can it be got rid of?

8. How are Bees made use of by man?

Chapter 9

1. What is peculiar about a Flat Fish?

2. A Frog's skin must always be moist. Why is this so?

3. How can you tell a Frog from a Toad?

4. Why are there few Reptiles in England?

5. Name three Birds that cannot fly. How do they get away from their enemies?

6. Why is there a keel on the breast bone of a Bird?

7. What Bird never makes a nest of its own?

8. What is meant by (a) Migration, (b) Hibernation?

9. How does a Kangaroo protect its young?

10. How do Rabbits warn one another that danger is near?

11. Although many Rabbits are killed by Man and other animals, yet there are still many Rabbits to be found in the country. Why is this so?

12. Why are Mice and Rats called pests?

13. From what animal do we obtain the best tortoiseshell?

14. Bats are seen flying about at dusk, and not during the daytime. How do they avoid hitting objects when flying?

15. How do we know that animals were living in the world many eras before Man existed?

Chapter 10

1. What are the regions of the spinal column, and how many bones are there in each?

2. What are the differences between the bones of a child and those of an adult?

3. What are (a) voluntary, (b) involuntary muscles? Give examples of each.

4. Compare the teeth of a child six years old with those of an adult.

5. What are "wisdom" teeth? Why are they so called?

6. Where is bile formed? Of what use is it?

7. What do we mean when we say that a particle of food has gone "the wrong way"?

8. Where is the bladder and of what use is it?

9. What are the special sense organs of the body?

CHAPTER 11

1. Some people do not use handkerchiefs unless they have a cold. Is this wise? Give reasons for your answer.

2. Do you think that sun bathing is a good thing or not? Give reasons.

3. Why is it so important to keep your teeth clean and healthy?

4. What are (a) infectious, (b) contagious diseases? Give examples of each.

5. What are the most common infectious diseases amongst children?

6. How could you tell that a patient was suffering from diphtheria and not scarlet fever?

7. Why is measles a dangerous disease?

8. What would you do to a patient who had fainted?

9. How would you treat a gnat bite?

10. If a person sprained his ankle, what would you do to relieve the pain?

11. How could you make a person sick?

12. What are the great dangers of a scald or burn?

13. Why do you think it is unwise to take "medicine" for constipation too frequently?

CHAPTER 12

1. What must happen to an egg before it can develop?

2. What living cells do not die?

3. How does a Hen's egg differ from that of a Frog?

4. How does a Kangaroo take care of its young?

5. The Sticklebacks are good parents. Why?

6. How do Ichneumon flies take care of their young?

7. Although a pair of Flies can produce about 20,000 maggots their numbers do not increase. Why is this so?

CHAPTER 13

1. How do we know that animals and plants lived on the earth millions of years ago?

2. Compare a Mammoth with a present-day Elephant.

3. Compare Pterodactyls and Bats with Birds.

4. Which are the older forms of plant life, Flowerless Plants or Flowering Plants? How do we know this?

5. In what order did the different groups of animals appear?

6. Say anything you can about (a) Charles Darwin, (b) Mendel.

Appendix B

APPARATUS AND MATERIALS NECESSARY

The following is a list of all the material required.

All the prices accompanying the apparatus specified have been taken from the catalogue of Philip Harris and Co., Ltd., Birmingham.

[The amount of apparatus required depends on the nature of the practical work, i.e. whether it is merely demonstration work

or whether it is individual work done by the pupils themselves.
The minimum amount is given below.]

1 microscope 3310 (recommended)	77s. 6d.
1 gross microscope slides. Ordinary quality	4s. 6d. per gross.
½ oz. cover-slips (round)	3s. 6d. oz.
Magnifier triple lenses (1 between 2 pupils)	2s. each.
3 filter funnels. Diam. 9 cm.	10d. each.
Filter papers. 12·5 cm.	1s. per 100.
1 measuring jar. 200 c.c.	1s. 10d.
6 gas jars. Diam. 5 cm., ht. 20 cm.	1s. 1d. each.
6 gas jar covers. Ground one side	3d. each.
3 iron wire gauze squares	3d. each.
1 Bunsen burner	1s. 3d.
1 tripod stand	1s. 3d.
1 doz. test tubes 5 × ¾ in.	7s. per gross.
½ gross assorted corks	7s. 6d. per gross.
2 bell jars. Diam. 14 cm., ht. 28 cm.	7s. 6d. each.
2 flasks. 375 c.c.	8d. each.
1 funnel with stopcock. 100 c.c.	4s.
1 batswing burner	1s. 6d.
1 lb. assorted glass tubing. Diam. 3–16 mm.	1s. 6d.
Specimen boxes [for keeping dry insects].	
All sizes. Size 3½ × 2½ × 1/12 in.	5s. per doz.
Specimen tubes with corks. All sizes. Size	
2 × 1 in.	1s. 4d. per doz.
Size 4 × 1 in.	1s. 8d. per doz.
1 lb. formaldehyde. 44 per cent.	1s. 2d. lb.
4 oz. chloroform	10d. oz.
1 oz. cobalt chloride	5d. oz.
1 lb. alcohol	1s. 10d. lb.
Iodine.	
Vaseline.	
Cotton Wool.	

Aquaria. Price depends on size. Disused accumulator jars can be bought very cheaply.

Jam jars, and any glass jars with screw tops can be collected by the children and are very useful.

Lime water. Shake up 1 teaspoonful of lime in 1 pint of water. Allow it to settle and then pour off the clear liquid. This is the lime water.

Prepared slides

Hydra.
Hydra budding.
Hydra with swallowed animal.
Body louse.
Head louse.
Amoeba.
Gnat male.
Gnat female.
Bed bug.
Tongue of bee.
Tongue of blowfly.
Tongue of butterfly.
Tongue of house fly.
Mouth organs of flea.
Mouth organs of gnat.
Mouth organs of garden spider.
Foot of house fly.
Sting of bee.
Sting of wasp.
Wing of bee hooked.
Wing of bee unhooked.
Spiracle of blowfly.
Trachea of blowfly.

1s. 6d. each or
15s. a dozen.

Specimens. Should be collected by the children as far as possible.

Appendix C

SOME USEFUL REFERENCE BOOKS

Batten, H. Mortimer. *Our Garden Birds.*
Bentham and Hooker. *A British Flora.* 2 vols.
Boulenger, E. G. *The Aquarium Book.*
—— *Animals in the Zoo.*
—— *Zoo Cavalcade.*
Cantlie, Col. Sir James. *First Aid to the Injured.*
Coward, T. A. *Life of the Wayside and Woodland.* 2 vols.
Daylish, E. Titch. *Name this Bird.*
Duncan, F. Martin. *Cassell's Natural History.*
Ees, Richard South. *Moths of the British Isles.* 2 vols.
Foster, Sir M. and Shore, L. F. *Physiology for Beginners.*
Fox, H. Munro. *Biology.*
Furneaux, W. *Human Physiology.*
—— *Life in Ponds and Streams.*
—— *The Sea Shore.*
Holmes, E. J. and Gibbs, R. D. *A Modern Biology.*
Huxley, T. H. *Lessons in Elementary Physiology.*
Jenkins, J. Travis. *Fishes of the British Isles.*
Johns, Rev. C. A. *Flowers of the Field.*
—— *British Trees.*
Joy, Norman H. *British Beetles. Their homes and habits.*
Lulham, R. *An Introduction to Zoology through Nature Study.*
Rolfe, R. T. and F. W. *The Romance of the Fungus World.*
Sidebotham, H. *Wild Animals.*
Step, Edward. *Toadstools and Mushrooms of the Countryside.*
—— *Wild Flowers Month by Month. In their Natural Haunts.*
    2 vols.
—— *Wayside and Woodland Blossoms.*
Thompson, H. S. *How to collect and dry flowering plants and ferns.*
Tinn, Frank. *Eggs and Nests of British Birds.*

Shown to the Children Series:

Billinghurst, P. J. *Beasts*.

Blaikie, A. H. *Nests and Eggs*.

Kelman, J. H. *Butterflies and Moths*.

—— *Bees*.

Scott, M. K. C. *Birds*.

Step, Edward. *Bees, Wasps, Ants and Allied Insects of the British Isles*.

—— *British Insect Life*.

Wyss, C. von. *Living Creatures*.

Appendix D

HOW TO FIND AND KEEP ANIMALS

Amoebae can be bought from dealers, or cultured in water in which grains of corn have been allowed to decay.

Hydra can be bought from dealers although it can be found (see p. 63). It can be kept in an aquarium, but it must be fed occasionally on Daphnia. Add pond water to tank carefully. Tap water kills the Hydra.

Earthworms. In a dry season if worms cannot be obtained, watering a patch of grass thoroughly, or stamping on the ground, will bring worms to the surface.

Leeches can be found in ponds, streams or canals and will feed on fish, tadpoles or worms. Cover the top of the aquarium or Leeches will get out.

Woodlice are found under damp or rotting wood.

Water Snails are plentiful in most ponds, streams and canals, and can also be bought from most local live-stock dealers.

Land Snails and **Slugs** can be found in the country, but are difficult to find in towns.

Insects

Butterflies and Moths. These are easily kept in school if the caterpillars are fed on the proper food. Caterpillars will not eat any plant, but only those on which they are found.

Silkworms can be bought from dealers and will eat Lettuce if no Mulberry is available.

Blowfly larvae can be bought from any shop that sells fishing tackle. Stick Insects can be bought from dealers. They are easy to keep, feeding only on privet.

Other Insects

Insects are easily kept if fed on the proper food. Details of insects not mentioned in this book will be found in the Reference Books.

Insect larvae living in water will live in an aquarium, if this is properly set up.

Preserving Specimens

Large insects, such as Beetles, can be kept in a 2 per cent. solution of formaldehyde, which also kills them.

Smaller insects, such as Flies and Butterflies, can be killed by putting them into a killing bottle containing cotton wool soaked in chloroform. A dry piece of cotton wool should be put on top so that the insect remains dry. When the insect is dead, spread out its wings before it stiffens and place it in a glass-topped box to keep off the dust. Butterflies and Moths should not be exposed to the light or their colour will fade.

The Aquarium

Put a mixture of silver sand and medium-sized stones at the bottom of the aquarium and plant a selection of water plants. [People in town can buy these from dealers.] If the plants and animals are well balanced the water should not need changing. Should the water need changing, siphon water in and out at the same time, if possible. If not, take out a few jars of water, and replace with clean water. It may be necessary, occasionally, to empty the aquarium and wash the silver sand; afterwards replacing the sand and weeds.

Fishes. Several Fishes are easily kept in an Aquarium. Goldfish can be obtained from dealers. Sticklebacks can be caught in ponds and streams and will feed on Tadpoles, Worms, or any water animals.

Fishes that live in running water are difficult to keep unless the aquarium has a constant supply of fresh water.

Amphibians. The eggs of all Amphibians can be obtained from ponds, canals or slow-running streams in March (see Expt. 4, Chap. 9).

All adult forms can be fed on Worms.

Reptiles. Lizards, Grass Snakes and Tortoises can be kept but require suitable cages.

Birds. Familiar cage Birds can be kept with little difficulty.

Mammals. Familiar pets such as Rabbits and Mice are very easily kept, but they require suitable cages or pens and need constant attention.

Dealers will give advice on how to rear the last three groups of animals.

Dealers in specimens of livestock

Names of dealers may be obtained from the journal of the "School Nature Study Union". This is issued to non-members for 1s. and can be obtained from the Secretary, 45 Cheviot Road, West Norwood, London, S.E. 27.

Index